Astronomers' Universe

For further volumes:
http://www.springer.com/series/6960

Robert W. Phillips

Grappling with Gravity

How Will Life Adapt to Living in Space?

Robert W. Phillips
Fort Collins, CO 80524, USA
rwpnbp@q.com

Please note that additional material for this book can be downloaded from http://extras.springer.com

ISSN 1614-659X
ISBN 978-1-4419-6898-2 e-ISBN 978-1-4419-6899-9
DOI 10.1007/978-1-4419-6899-9
Springer New York Dordrecht Heidelberg London

Library of Congress Control Number: 2011937452

Printed on acid-free paper

Springer is part of Springer Science+Business Media (www.springer.com)

To Nancy,
For everything.
Without her support and help
this book would not exist.

Acknowledgements

Without the astronauts/cosmonauts and their dedicated work in space, this book could not have been written. My wife Nancy has made innumerable corrections and suggestions while repeatedly reading drafts of the manuscript. In addition to her input, the greatest single contributor to the book is F. D. (Denny) Giddings, who transformed my rough concepts into beautiful and illustrative artwork that can be found throughout the book. Charlie Kerlee's and Jay Oats' skill in editing videos was invaluable.

I greatly appreciate the input from NASA and other colleagues who have proofed the manuscript and made many worthwhile suggestions. Joan Vernikos and Tom Dreschel both made many helpful comments and corrections. Gary Coulter provided a great deal of help in the chapter on immunology. Mike Hand suggested Chapter two. Doug Parks labored through the manuscript with me as he helped clarify my thoughts. Al Billington provided helpful advice. Our son Lex supplied a much needed layperson's approach and his son Chris made several valuable suggestions. Our daughter Logan and her husband Tim Mather contributed to several portions of the book. I cannot name individually all of the scientists and researchers whose findings on space-induced changes are the basis for the book. They are the ones who designed and flew countless biomedical experiments. I hope that I have correctly translated their findings. The videos that are so illustrative are important to the overall appreciation of what happens in space. Kenichi Ijiri's videos of breeding fish and Richard Simonds' copy of the NASA video of adaptation of fish on Skylab are important as is the quail chick video from Mir provided by colleagues at the Institute of Biomedical Problems in Russia. Ken Souza of NASA allowed me to use his pictures of tadpoles during parabolic flight. Other videos are drawn from NASA archives and spaceflight experiments depicting changes in behavior while in space or upon returning. Julie Piotrascke's editing has greatly improved the flow of the chapters and the book. I am deeply indebted to all of these friends and colleagues.

Contents

About the Author

Robert W. Phillips is a veterinarian and has also a doctorate in Physiology/Nutrition. He taught and conducted research at Colorado State University until 1984, when he was selected by NASA to fly on "Spacelab Life Sciences 1," the first shuttle mission devoted to biomedical research. Ten months prior to launch Phillips was removed from flight status due to a medical condition. He supported the mission from the payload operations control center. Following the mission, Phillips served for 3 years at NASA HQ as the Space Station Chief Scientist during the station's design phase. He then joined the NASA Life Sciences Education and Outreach program for 10 years, presenting seminars and workshops to teachers, students, and the public about the effects of spaceflight on life.

Part I
Space Here We Come

1. A Great New Adventure

I shall be telling this with a sigh
Somewhere ages and ages hence:
Two roads diverged in a wood, and I—
I took the one less traveled by,
And that has made all the difference.

The Road Not Taken, Robert Frost (1873–1964)

There are many choices to be made in life, decisions that can change our whole existence. A decision you may have to make will be whether or not you will go into space. Our ancestors chose to go to new states, countries, even continents. For future generations the choice will be new worlds. This book is about living in space. It explains how to get there, what to expect, how it will change you physically, and more. Perhaps at first you will be a space tourist. Then a bit further into the future, you will choose to live off of Earth permanently. Maybe you will live on the moon or Mars or in a space colony. These opportunities are almost here.

Living in space requires changing and adapting to new environments. Freeman Dyson, visionary physicist and mathematician said, "surviving and making a home away from Earth are problems of biology rather than engineering (*The Sun, the Genome and the Internet*, Oxford University Press, 2000.)" The fact that we are part of the problem doesn't prevent us from providing the solutions. There is no reason to believe that Earth life needs to be contained on Earth. We have both the capacity and the desire to turn other worlds into a new home for the human race. Nobody says it will be easy, but it is likely to happen.

One thing before you launch on this great trip: you need to like people and togetherness. If that's not for you, better to become a recluse if you are looking for a chance in life style. For the rest of us, it is a great adventure: the first new one to come along in quite a while. If you just choose not to go, that's okay too. Leaving Europe to cross the Atlantic to a new world wasn't for everyone

R.W. Phillips, *Grappling with Gravity: How Will Life Adapt to Living in Space?*, Astronomers' Universe, DOI 10.1007/978-1-4419-6899-9_1,

either. Somebody has to keep the home fires burning and that would be most of the population. As always, it will be the lucky, adventurous, or perhaps crazy few who will make the trips that fuel the future.

Until now, space travel has been limited to a number of select, highly trained astronauts, or the very rich. These days are nearly over. Average citizens will have the opportunity to journey away from Earth's surface and share this experience with the previously select few. Eventually space flight will not consist of just a brief visit to a strange new environment, but will entail living there, raising a family, and perhaps become a new race of humans adapting to the new environment. Before very long it will become possible to live in space or on neighboring worlds such as the moon or Mars. Once we have conquered these nearby destinations, we will undoubtedly move out farther. We may go to other parts of our Solar System or much further in the future, to planets around other stars.

Let's now explore some frequently asked questions about space travel in the near future, such as: "How far up is space?", "How high do I have to go to become an astronaut?", and, "what kinds of trips are likely to be offered in my lifetime?"

1. Suborbital Trips: On a suborbital trip, you travel 100 km (62 miles) above the surface to reach space. The whole trip will take about an hour or even less. If you don't get that high you have not been in space. A suborbital trip is perfect for those who just want to claim the "astronaut" title or briefly view Mother Earth from a new perspective. They don't necessarily care about the other benefits of a space visit or a prolonged stay.
2. Low Earth Orbit (LEO): Now you are 150 or more miles up, going forward at 17,500 mph, with really great views of our planet. It takes just 90 min to circle Earth at that speed. Spend a couple of days enjoying the ever-changing scenery.
3. Space Station Visits: You will still be in LEO, but probably a bit higher, with more creature comforts and a better opportunity to view the unfolding panorama of Earth.
4. Elliptical Orbit: At the apogee, or highest part of the ellipse, you will be much farther from Earth, perhaps several thousand miles, and can begin to visualize our planet as a single splendid unit, floating all alone in space, the "Blue Marble."

5. Off to the moon!: You may go either as a tourist or perhaps as one of those hardy pioneers out to start a new life, committed to turning the moon into another habitable world for the human race.
6. Destination Mars: Mars is very different from the moon, much larger and with a stronger gravitational pull. The trip to Mars from Earth takes about 6 months, compared to only a couple of days to reach the moon.

After a while, people who live in space, on the moon, or on Mars will be different than the Earth dwellers they left at home. You need to know what will happen to your body without the strong and continuous gravitational pull from Earth? We are just starting to learn about that now, particularly the effects of long-term trips away from Earth. All of the changes can be lumped into *adaptation.* So far there has not been a single person who has spent even 1/50th of his or her life span in space. There is a lot of conjecture and speculation about long-term space flight adaptations, but only a smattering of information. We are still fumbling in the dark. Our goal is to get out of our visitor phase and think about living in a new environment.

Before we consider what happens when we make a permanent move into space, let's explore what happens on just a short trip, such as those trips taken by astronauts today. To start are the suborbital flights to be developed by commercial organizations. These flights are planned to be brief excursions of a few minutes at or above the 100 km mark. They are called suborbital because the vehicle is not going forward at a speed fast enough to stay in space. It is sort of like shooting an arrow into the air and watching it fall back down. In this case, the spaceship will either parachute back after it has re-entered the atmosphere or glide back to a landing near the flight's origin.

It's possible to reach that 100-km goal with a combination of a large airplane and a small rocket ship. The airplane carries the rocket ship up 10 miles above the thick lower atmosphere and then drops it off. The engines ignite, and the rocket ship blasts the rest of the way up to or beyond that 100 km mark. It is a great thrill, but you can't really appreciate the view of our planet as seen from space in a few minutes of exposure. Imagine trying to experience a movie by only viewing the preview, tour the Smithsonian Institution in Washington, D.C. in 20 min, or experience the majesty of the Grand Canyon for 30 s. It can't be done. Those first astronaut-qualifying flights won't last long enough for the participants to enjoy some of the benefits of a longer stay. They won't begin to adapt if the trip is simply up and back.

However, these early flights will be a prelude to the real thing – going into orbit. Low earth orbit (LEO) is 150–500 miles above the surface and is high enough so that there is really no atmosphere. The closer the vehicle is to Earth's surface, the thicker the atmosphere, and the greater the drag slowing the vehicle. The spaceship has to get high enough to be above the atmosphere and fly forward at 17,500 mph. LEO is the easiest space destination to reach. However objects in LEO are not in a stable orbit, they will eventually re-enter Earth's atmosphere. The space shuttle usually flies about 200 miles up, and the International Space Station (ISS) is somewhat higher. The ISS is continually present in space, yet it slowly sinks towards Earth. Because of this, it must be re-boosted back to a higher altitude every couple of months. The higher it is, the less frequently it needs another boost.

Like any trip you take on Earth, you need to know how you are going to travel. Today, the cost of space transport is huge, and uses large chemical rockets that are currently available. These chemical rockets started our travel to space, and we have not gotten out of that phase. As technological improvements in space travel become available, a whole new industry will jump-start. Most tourist trips will be at the lower end of LEO, about 200 miles up. This is roughly the distance from New York to Boston, or Miami to Orlando when driving on Earth's surface.

As we have begun to travel or send probes to other worlds, it is obvious that Earth is the oasis around the Sun, but that won't be proven until humans actually explore Mars and look under the ice of Jupiter's moon, Europa, to see if life is present in that ocean world with its permanent cover of ice. Our planet, a large blue sphere, is a magnet that attracts us to return home when we venture into space. It would also surely attract travelers from other worlds if they happened to pass this way as they explore new solar systems (Figure 1.1).

FIGURE 1.1 Earth as seen on the way to the moon. North America is mostly clear, but some clouds and weather are approaching from the northwest. With this picture's resolution it would appear to be a vacant and desolate world. The only visual sign of life is the *green* color of chlorophyll in plants. There is no evidence of advanced or intelligent life. (Picture courtesy of NASA)

Just as our ancestors – oasis dwellers and villagers – left their comfort zone to explore other parts of this planet, we search for other worlds that are, or can be, new oases for us. Astronomers are looking for Earth-like planets orbiting nearby stars that are similar to the Sun. When one is found, the question will be, should we reach out, should we spread Earth life to other worlds that are so far away? Even the thought of such a voyage seems impossible given today's technology or even our best guesses on the technology of the future. Table 1.1 shows the relative distances from Earth to the Moon, the shortest distance to Mars, the distance to the nearest star, the Sun, the distance to Alpha Centauri (a close neighboring star), and the distance across our galaxy, the Milky Way.

TABLE 1.1 Distances from Earth to other destinations. Space travel outside of our Solar System is a frequent theme of science fiction, but it is an unlikely scenario even if we could travel at 1/100th of the speed of light, or 6,700,000 mph. At such a speed, it would still take 430 years to reach the vicinity of the nearby star Alpha Centauri. Sun-like stars are farther away. In comparison, the space shuttle and ISS circle Earth at 17,500 mph in just 90 min. To escape Earth's gravity and go to the moon, you need to travel over 25,000 mph. We're not even close to leaving our little corner of the Solar System

SPACE DISTANCES FROM EARTH	
• Low Earth Orbit (LEO)	150-500 miles
• Moon	238,000 miles
• Mars	40,000,000 miles
• Sun	93,000,000 miles
• Alpha Centauri	25,000,000,000,000 miles or 4.3 light years
• Across Milky Way Galaxy	100,000 light years

The nearest Sun-like star is about four times as far away as Alpha Centauri. Is it any wonder that our little planet has not been – as far as we know – visited by voyagers from other worlds? UFOs, alien abductions, and strange encounters of whatever kind are fun to talk about as fantasies, but they really have no place in the discussion here. So far, in spite of considerable searching, we have not yet detected radio signals from distant civilizations.

Advanced civilizations have come and gone on this planet for thousands of years, starting with Egypt, then China. Over 8,000 years of written history, yet only now do we have the technology to broadcast our existence towards distant, unknown civilizations. The first long radio broadcast from Earth occurred in 1906. The information in that electronic signal is still traveling out from Earth at the speed of light and is about 100 light years away. If anyone out there is listening, they would have to be close, on a galactic basis speaking. Given the 100,000 light years that span our Milky Way Galaxy, we haven't advertised our presence very broadly. If that signal was received by another technically advanced society, and we were expecting a response, it would only be half that distance, about 50 light years away. For the foreseeable future, we are on our own.

Traveling closer to home, the first trip to Mars with humans aboard will take about 6 months. It is only 40 million miles away, practically next door. Traveling at the same speed we'll use on that Mars trip it would take about 300,000 years to reach the vicinity of Alpha Centauri, our close galactic neighbor. Talk about impossible assumptions or flights of fancy. We can't even consider a trip to the vicinity of Alpha Centauri or any other sun with the technology available to us at this time.

One of our great advantages in becoming spacefarers is the closeness of our moon. Not all planets have a large, close satellite with available resources just next door. An Earth-moon shuttle can deliver people and cargo in only 1–2 days. That can be a lot easier to manage than the situation faced by the "European settlers of the Americas."

In the last few decades, space flight has brought a new dimension into our lives. Since life began on planet Earth hundreds of millions of years ago, no organism had ever experienced the excitement of space flight until now. The small dog Laika was the first in 1957. Space flight has opened the door to the future. Until now life has been confined in the relentless grasp of Earth's gravitational field. You can't change it or ignore it. You have to learn to live with it. Life has done a pretty good job of that. We are not blobs lying flat on the ground. We have evolved to challenge gravity. Plants grow up against the gravity vector by using thick cell walls that provide support. We and many other vertebrates stand erect and move about. But still, gravity controls how we look and how we act. Long-term space flight will bring big changes in our bodies. Knowledge of such changes will allow us to make some predictions of what may occur when we leave Earth and return to the moon to visit and live, and then move outward to Mars on a permanent basis.

So far, travels away from our oasis have just been brief camping trips (see Figure 1.2). We carry everything with us. We have either traveled just outside the atmosphere for brief periods or

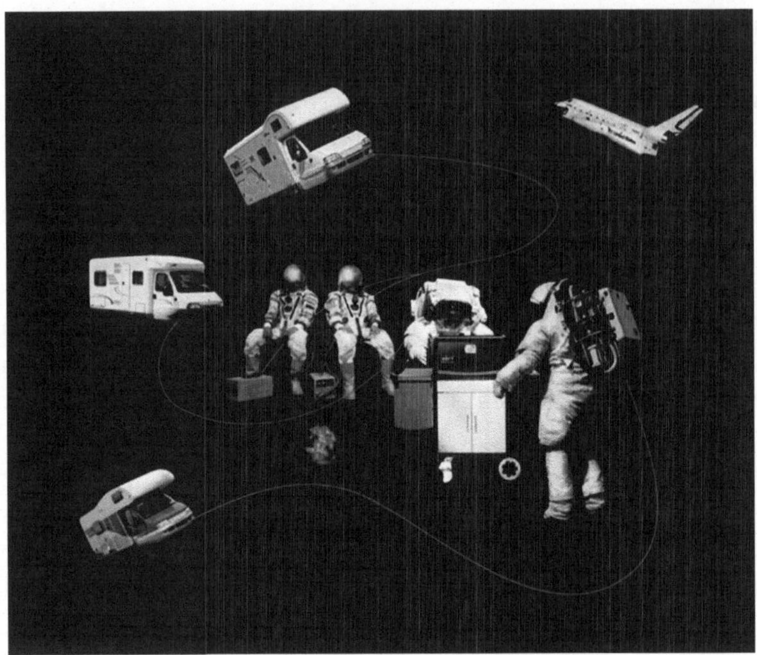

FIGURE 1.2 Right now, space trips are like camping trips on Earth. We carry everything that we need – housing, food, water. We will begin to truly live away from Earth only when we have built living facilities in space or on other worlds, grown food, and become self-sufficient. Outposts and colonies are the natural next steps in human expansion to new worlds.

taken short trips to our moon. Today, a mission may be a couple of weeks on the shuttle or a number of months on the ISS. In earlier days, when the Russian MIR space station was in orbit, several cosmonauts spent a year or longer in space on one trip. That is the record at this time.

A year in space is a small fraction of a human life, not long enough to become completely adapted to the new space environment. All supplies are provided as needed from Earth. We have a lot to do before we reach that time when we are living away from our planet, producing our own food and recycling all wastes. Because the moon is only a day or two away, it will be possible to re-supply lunar outposts while the explorers and colonists learn how to become self sufficient. Acquiring this background knowledge and expertise relatively close to home will be a big step in the process of exploring and then settling Mars.

As we venture more and more into space, and for longer periods, we are going to find out how we, other animals, and plants respond to this great environmental change. Some changes will be good, some bad, and some unknown. The good is easily appreciated. We change, we adapt, we go with the flow and learn to function in this new situation very rapidly. Not so for very young animals, particularly mammals such as we are, and also birds. This is a possibly serious problem that we haven't even started to solve. (More on this later.) For the rest of us, all it takes is an earthlike atmosphere and temperature environment. These can be available inside a space vehicle or habitat on some other planetary surface. Food, water, and waste handling are also necessary. The bad news is that, as we adapt to the space environment, we become less adapted to Earth's gravity. If we spend a while adapting to space, then we have to relearn the art of living in a 1 g gravitational field when we return home. The unknown is the question of what happens to generations of space travelers who have never been exposed to 1 g but instead live throughout their lives in space or a decreased gravitational force. How will their bodies change? How will they behave? We know enough to make some predictions, but there are also undoubtedly many surprises in store.

Many books have been written about human space flight. Most have focused on the exploration, excitement, or thrill aspects, or on the engineering feats needed to launch spacecraft and to support life in space. This book describes how flight in space affects Earth life. We don't understand all of the long-term problems. We need to learn how years of space flight, or living in colonies on the moon or Mars, may change life forms that originated on our planet. We need to know what problems must be overcome in order to prosper during a lifetime in space, or in a lunar or Martian outpost or colony.

Right now, we are developing the capability to return to the moon and establish a permanent outpost. That will be the beginning of living away from Earth. It will lead to human flights, and then outposts and colonies on Mars. Many countries, as well as commercial entities, are interested in achieving a permanent human presence away from Earth, and have the moon and Mars as destinations. The next breed of explorers and settlers will doubtless be multinational. They will need to be committed to the mission for all of the citizens of Earth, not for the glory or good of just one nation. The urge to explore is deeply ingrained in our psyches. This fundamental component of our being has led us out of Africa to populate our planet Earth. It is time to turn it loose again.

Neither the moon nor Mars can sustain life as we know it without an enclosed habitat because of severe temperatures and the lack of a breathable atmosphere. We must take much of our "Earth" environment with us as we begin to move out. The solution is straightforward, and we have the technology to do it. It is not really different in concept from the notion of living under the sea in a submarine, or a permanent aquatic habitat such as Aquarius, that is in operation, or Atlantica, which is being developed. Submariners have been supplied with an earthlike atmosphere for years. However, submarines and underwater habitats are still in an Earth gravity environment. That is a big difference.

We can create pressurized habitats almost anywhere with a temperature range and atmospheric composition similar to Earth. What we can't do is create living quarters with an Earth normal gravitational field. Centrifugation can produce a gravitational field, but we have yet to seriously plan for large, habitable centrifuges in space. The expense of such a project is far beyond our budget.

The closest solid body to Earth is our rocky moon. The moon's gravitational force is about one-sixth as strong as Earth's. This means that a person will only weigh one-sixth as much while on the moon. To determine weight anywhere all you have to do is multiply the force of gravity by body mass, as shown in Figure 1.3. A person weighing 150 pounds on Earth would only weigh 25 pounds on the moon. What an opportunity that will be when we have moon habitats with room to maneuver! All that is needed is a large enough space so that your Earth muscles can take advantage of the decrease in weight. A new ball game might include a slam-dunk 30 ft in the air. There would be plenty of time after the dunk for some great flying gymnastic feats while you were slowly falling to a soft landing on the surface. What a blast, at least until you get to be an old timer with moon-adapted muscles.

We are not ready to do all of these things. The technology isn't quite there yet that could fulfill our dreams and aspirations. However, having something to strive for is what brings about innovation and invention. What we need most is to replace chemical rockets and let some of the new plans and ideas about going

FIGURE 1.3 If you weigh 150 pounds on Earth you will weigh only 25 pounds on the moon, as its gravitational force is only 0.16 of Earth's gravitational force. Body mass does not change, only weight. It will still be you, but with much greater physical ability.

to space mature. It is happening now. Read on. Come aboard and join the future of our children and grandchildren as they move out from their planet of origin. The sky is dark and deep and the human race has many miles to go to become spacefarers, distances that would be unimaginable to our parents and grandparents. Our promises are to the memory of those ancients who strode out of Africa, and to the generations yet to come.

2. The Long Road to Space Flight

LISTEN TO THE MUSTN'TS

Listen to the MUSTN'TS, child,
Listen to the DON'T'S
Listen to the SHOULDN'TS
The IMPOSSIBLE, the WON'TS
Listen to the NEVER HAVES
Then listen close to me---
Anything can happen, child,
ANYTHING can be.

Shel Silverstein (1930–1999), Poet

For thousands of years, our ancestors have dreamed of leaving Earth and traveling to space. The Greeks told one of the first stories regarding a trip into the sky. As the tale goes, Icarus and his father Daedulus planned to escape imprisonment on the Island of Crete by fashioning wings of wax and feathers. Icarus and Daedulus were successful and flew into the sky on a bright sunny day. Icarus was so excited with this new freedom that he flew too close to the Sun and melted the wax that held his wings together. He fell into the sea and drowned (Figure 2.1).

The early Greek dream of human flight has been passed down through generations. Early in 1010 the theory of human-powered flight was tested in England. A monk, Eilmer by name, built wings and fastened them to his arms and feet hoping to repeat what he believed was Daedulus's great adventure. He launched from the top of the old Malmesbury Abbey or a nearby Saxon watchtower in England. The towers were both about 80 m tall. He managed to glide for about 200 m before losing control and crashing to the ground. He broke both of his legs in the crash and walked with a limp for the rest of his life. Eilmer was not discouraged by the accident, as his superiors had to forbid him from making a planned second flight. His glider is the first recorded example of human

R.W. Phillips, *Grappling with Gravity: How Will Life Adapt to Living in Space?*, Astronomers' Universe, DOI 10.1007/978-1-4419-6899-9_2,

FIGURE 2.1 According to an ancient Greek myth, Daedulus and his son Icarus attempted to escape a prison on the Island of Crete. They built wax wings with feathers stuck in the wax and flew into the sky. Icarus, in spite of parental warning, flew too close to the sun which melted the wax. He fell into the sea and perished.

flight, even though it ended with unfortunate consequences. The current Malmesbury Abbey has a stained glass image of Eilmer that celebrates his ingenuity and bravery (Figure 2.2).

In the fifteenth century, Leonardo da Vinci sketched images and designs of gliders, helicopters, and ornithopters all devoted to the concept of human flight. His human-powered flapping-wing ornithopters have never come to be. Our arms and chest muscles are not capable of powering flight. However, gliders and helicopters are part of our modern day culture. Da Vinci's helicopter design

FIGURE 2.2 Eilmer of Malmesbury, a monk in England, is generally given credit for being the first human to fly. A likeness of Eilmer adorns the current Malmesbury Abbey located in Wiltshire, Southwest England. (Photo courtesy of Andrew Dunn)

was patterned after a child's toy that had been imported to Italy from Asia in the previous century.

The first flight from Earth was almost 300 years after Da Vinci drew designs of various flying machines. In France, two Montgolfier brothers devised a balloon that would be able to carry people into the sky. Their design was based on observations by one of the brothers, Joseph, who was something of a maverick. He had noted how clothes drying over a fire billowed upwards and believed that the smoke contained a special gas that provided lightness or levity. In 1782 he built a box of light wood, open on one side, and covered it with taffeta cloth. He lit a fire and placed the box over the fire with the open side down. As the air inside the box heated and expanded, the box lifted and hit the ceiling. That was the beginning. The brothers conducted a number of experiments with larger and larger balloons.

On September 19, 1783, at Versailles outside of Paris the brothers launched a balloon with passengers that flew about 1,500 ft into the air and landed safely almost 2 miles away (Figure 2.3). This first flight carried a rooster, a duck, and a sheep. The Mongolfiers felt

FIGURE 2.3 This is an artist's rendering of the first time that passengers ever left Earth's surface. For centuries, people had been dreaming about flight and envying the freedom of birds to casually fly and land.

that the duck would not be affected, as they normally fly in the sky, but roosters are ground birds and might somehow be changed. The sheep was included as they believed that sheep were similar to people. Just as space launches draw big crowds today, King Louis the XVI and Queen Marie Antoinette watched the trip. Several months later, humans began to venture off Earth's surface in the Montgolfier's balloons, but the brothers stayed on the ground and never experienced the magic of flight.

A little over 100 years later in the 1890s, Otto Lilienthal of Germany and Englishman Percy Pilcher developed and flew gliders. Both died in glider crashes. These early gliders were best launched from hilltops and could fly for short distances. Their flight experiences helped guide the Wright Brothers gliders they launched from Kill Devil Hill. They were working on building their successful flying machine powered by the new internal combustion engine. The first flight of a "heavier than air" flight was accomplished by Orville and Wilbur Wright on December 17, 1903, at Kitty Hawk on the Outer Banks of North Carolina. They made three flights that day each one longer than the one before it (Figure 2.4).

FIGURE 2.4 This picture shows the beginning of human powered flight. It is the Wright brothers' first flight at Kitty Hawk, North Carolina, on December 17, 1903.

Following the Wright brothers, airplane use expanded rapidly. In the early 1900s both automobile and airplane requirements for propulsion energy fueled the development of new, lighter, and more efficient gasoline engines. Airplanes were used extensively in the first World War. Many models of fighters and bombers were used by the competing armies.

A big step forward in aviation was taken with Charles Lindbergh's flight from New York City to Paris, France, in May 1927 in his small airplane named the *Spirit of Saint Louis* (Figure 2.5). It was the world's first solo intercontinental flight. It started a revolution by establishing air transport as the preferred method of travel linking cultures and societies around the globe. Air travel rapidly went from a few thousand passengers each year to hundreds of thousands of passengers per year

As airplanes became more sophisticated, larger, and faster, a new technology was being developed. Robert H. Goddard's experiments with rockets started in 1912, near his work at Clark University in Worchester, MA. His work would lead directly to rockets that could carry humans into space. On April 12, 1961 Yuri Gagarin, a Russian cosmonaut, was the first human to experience space flight. He was launched on a Vostok 1 rocket and made one orbit of Earth, landing safely by parachute in the Soviet Union. His flight was followed on May 5, 1961 by Alan Shepard of the

FIGURE 2.5 Charles Lindbergh's airplane, the Spirit of Saint Louis. Lindbergh flew alone from New York to Paris, France, in May 1927. This was the world's first solo non-stop trans-Atlantic flight and a milestone in the development of air travel.

United States. In his Mercury capsule named Freedom 7, Shepard flew a suborbital mission, landing hundreds of miles downrange from his launch site. He was followed a number of months later by the American Astronaut John Glenn, who was launched in a Mercury capsule named Friendship 7 on February 20, 1962. Glenn made three orbits of Earth before parachuting into the Pacific Ocean (Figure 2.6). By today's standards, these early space flights were in small, simple space vehicles. There was little room to move about.

FIGURE 2.6 Launch of the Mercury capsule Friendship 7 on February 2, 1962. It carried John Glenn around Earth three times. It was the first orbital flight of an American, although the Russians had several earlier orbital flights that carried people.

Early space vehicles were built to be partially destroyed during re-entry into the atmosphere. They had a shield on the forward surface that partially melted away as friction with the atmosphere heated the spacecraft. New innovations provided a reusable spacecraft, the space shuttle, that was not generally damaged during re-entry. It was designed to launch like a rocket and land like an airplane. The first shuttle mission using the space shuttle Columbia launched from Kennedy Space Center Florida on April 12, 1981 and landed on a runway 2 days later at Edwards Air Force Base in California (Figure 2.7). It carried a crew of two astronauts, John Young and Robert Crippen. The Columbia launch was exactly 20 years after Yuri Gagarin's first flight into space.

The next big space venture is to return to the moon to stay. Currently Russia, China, India, and the European Space Agency are working towards establishing permanent moon bases.

FIGURE 2.7 Columbia ready for launch at the Kennedy Space Center, April 1981. The astronauts used their time in space to check out the shuttle's systems. (Picture courtesy of NASA)

The United States was removed from the return to the moon agenda in 2010 by President Obama. NASA's new Ares lunar rocket is no longer being developed.

Today, there are limited opportunities to go into space. The first real chance for ordinary people to make suborbital trips is coming soon, just up and back; a quick excursion easily done in an afternoon. Suborbital space trips will be a brief phase, like barnstorming was in the early part of the last century at county fairs. Barnstorming gave people the chance to ride in an airplane for the first time. The pilot would fly to the fair, perform some aerial acrobatics, then land the plane and sell rides to eager onlookers.

During the Apollo program, six groups of three-man astronaut teams traveled to the moon, and two members of each team landed on the surface. This was an exciting time in space travel. No one had ever ventured away from our planet before. The Apollo pioneers proved that we could travel to, land, and then leave from a neighboring world. While on the moon's surface, they explored and deployed a number of scientific experiments. The explorers investigated the region immediately near their landing sites on foot and used the lunar rover for longer excursions. They collected and returned to Earth, for study, many samples of soil called regolith, or moon rocks. Several experiments proved that lunar soil could support plant growth in Earth's atmosphere.

When we think about how space flight developed, it helps to reflect on the progress we have made. It was several thousands of years from that ancient Greek dream about Icarus and Daedelus flying until Eilmar of Malmesbury made his short flight in 1010. About 400 years later, Leonardo drew designs of various flying craft, all gliders or human powered. In another 300 years, the Montgolfier brothers developed their balloons. One hundred twenty years after that, the Wright brothers' first airplane flew for a few hundred yards. Only 24 years more, and Charles Lindbergh flew nonstop from the United States to Europe in his small single-engine airplane. Gagarin's first trip into space orbiting Earth was another 34 years. Eight years after that, the first moon landing occurred when Neil Armstrong and Buzz Aldrin arrived safely on the moon and began exploring our closest neighbor in space. Technological change feeds upon itself. Each innovation or discovery speeds the development of the next new wonder. The old cliché 'one thing leads to another' is confirmed by the development of every new facet of air and space travel.

This great record of achievement involved not just those who flew, but the teams that supported them and the untold numbers who believed in what these pioneers were doing. They were certain that we could conquer flight from Earth and then flights into space. The next great venture is not only to visit but to live on other worlds. Many dedicated people are working today to make that happen.

3. Why Go to Space?

Exploratory space flight puts scientific ideas, scientific thinking and scientific vocabulary in the public eye. It elevates the general level of intellectual inquiry. The idea that we've now understood something never grasped by anyone who ever lived before—that exhilaration, especially intense for the scientists involved, but perceptible to nearly everyone—propagates through the society, bounces off walls and comes back at us. It encourages us to address problems in other fields that have also never been solved...people everywhere hunger to understand.

Pale Blue Dot, Carl Sagan (1934–1996)

Part A

People have been itching to leave Earth ever since we knew there were other places to go, that Earth was not the only show in town. It is our innate curiosity, the mystery of what is beyond the next hill or the moon or the next planet. For thousands of years humans have sought to understand the Sun, our fellow planets, and other stars. It is only now that we can, not just view, but travel to other worlds and we have done so. The last or perhaps next great adventure for the human race is to leave the only home we have ever known, Earth. It is a strong driving force and we love to solve mysteries.

Solving mysteries is not a new idea. Copernicus way back in 1543 and then Galileo in 1609 started us off. Galileo's telescope showed that Copernicus was right. The Earth went around the Sun. Everybody else just knew that we were the center of it all and that the Sun went around the Earth. This was radical thinking back in those days. Since Galileo would not recant his heresy, the Church locked him up. A few years ago, they finally admitted he was right,

R.W. Phillips, *Grappling with Gravity: How Will Life Adapt to Living in Space?*, Astronomers' Universe, DOI 10.1007/978-1-4419-6899-9_3,

but it was a bit late. He had been dead for almost 400 years. He was the start and we've been building on his success ever since. Our earthbound and space telescopes continually provide new and exciting information about other worlds. We use space missions to photograph other planets and moons and to analyze their atmospheres. We send crawling robots to travel across and sample the surface of the moon and Mars. The Apollo flights 40 years ago are the only human visits to other worlds. They were in brief excursions to our closest neighbor, the moon. Now we are headed back to the moon and at last, off to Mars. Eventually, we will visit other parts of the Solar System. Our new voyage of discovery and adventure should be personal, not by proxy. Robots can do a good job, but they can't substitute for the curiosity, imagination, and drive of people. Robots are definitely second best. Can you imagine a robot writing poetry about a Martian sunset or lunar earthrise? Of course not, man needs to go.

From the dawn of man, human transportation has progressed from walking to riding to boats, trains, cars, planes, and spaceships (Figure 3.1). Each new innovation in transportation has been developed to take advantage of technological advances. We sometime forget, or ignore, the fact that the simple dugout canoe was a precursor to ocean liners, or that the Wright brothers' airplane evolved into the biplane now replaced by hypersonic military planes and jumbo jets that are so common in our sky today. It is no more logical to assume that leaving Earth will always be accomplished by chemical rockets than it would have been to assume that the canoe and biplane would always be the means of water and air transport. Our civilization has been, and will be, advanced by often unanticipated new technologies. Mercury, Gemini, Apollo, and the Space Shuttle are the beginning, not the culmination, of spaceflight developments.

Two dreamed of technologies that could support space travel are not ready for use. They are controlled fusion and antimatter propulsion. The most promising fusion technology will utilize helium 3 that is present in relatively large quantities on the moon. Only small quantities are available on Earth. As we return to the moon and mine helium there, it can eventually be used in fusion reactors to provide an inexpensive and clean source of energy. No radioactivity and no carbon emissions are released from helium 3 fusion technology.

FIGURE 3.1 Our means of transportation have evolved greatly over the centuries. We have progressed from walking around our neighborhood to traveling in space. We can visit worlds that we were only able to dream of before the space age. Each mode of transportation shown is for the most part the beginning of a new technological revolution. The dugout canoe has evolved to become an ocean liner; the steam engine is now high speed intercity transport. The buggy that became a horseless carriage, then an automobile, aptly named because it represents auto or self mobility. Today there are hundreds of millions of automobiles around the world kept busy in our transport. The jumbo jet's precursor is the biplane that followed the Wright brothers. The Mercury Rocket and the Space Shuttle are the beginning, not the end, of the coming era of space flight.

Antimatter propulsion rests first on our ability to form larger quantities of antimatter at a reasonable cost. Antimatter protons have a negative charge instead of the positive charge found in protons common on Earth. When these two types of protons are combined, a tremendous amount of energy is released. A fraction of a gram of antimatter has more power than the total amount of fuel used to propel a space shuttle into orbit. It is almost unbelievable that a fraction of a gram, a mere smidgeon, has greater power than tons of combustible fuel. We know the usable energy is there, but we have yet to harness it. Both of these potential new systems can work, but a great deal of ingenuity and innovation has to be focused on the process to make either of these energy sources a practical solution.

It is possible that we will develop the capacity to move spacecraft at the speed of light. That would be over 11 million miles per minute! However, the nearest star, similar to the Sun that may have usable planets surrounding it, is 16 light years away. We are a long way from proposing such a trip, as it would take 32 years just to get there and back without any exploration. Radio communications would also take 16 years to reach Earth after being sent from the vicinity of that star. It would be hard to carry on much of a conversation in that time frame. You would undoubtedly forget the question you asked long before the answer arrived, and it would probably not be relevant to your children anyway!

Several less ambitious technologies are closer to reality. One proposed technique would use electromagnetic propulsion in a vacuum tunnel up the side of a high mountain. The spacecraft would exit the tunnel at thousands of miles per hour, above the denser part of the atmosphere. Rocket engines would ignite once the spacecraft was above the denser part of the atmosphere and provide the vehicle the speed necessary to go into orbit around the Earth. It would be a step in the direction of reducing the cost of access to space, but still would use some rocket fuel. Electromagnetic propulsion is very effective, but on short launch tracks the necessity for rapid acceleration would produce more g forces than humans can tolerate. No new technology is needed to build electromagnetic propulsion systems, just the decision to move forward. Several different technologies are being worked on to provide easier access to space. One, the space elevator is described later in the book. The elevator will carry passengers and cargo to space using laser or other future propulsion techniques. Using this technology the elevator would move along a cable, or highway, made from something called carbon nanotubes. Carbon nanotubes are a relatively newly recognized form of carbon and quite different from other types of carbon, such as soot, charcoal, and diamonds. The nanotubes are very small, yet unbelievably strong. Just as diamonds are the hardest material known, nanotubes are the strongest material known. A series of nanotubes connected together with a total diameter of 1 mm could support a weight of over 6,000 pounds. That strength is over 150 times as great as the strongest steel. One problem with a space elevator is that you will travel relatively slowly into space, only a few hundred miles per hour.

We have only begun to develop technologies to allow us to journey away from planet Earth. At this time, our only option is to use rockets. Rockets are the beginning, not the end of our quest to conquer space. The best is yet to come. Since the first space flight, the adventure of going into space has been a lure attracting many bright and innovative young people into this exciting field. It has been a stimulus for education in science and technology. This impetus will continue and strengthen as new opportunities to work and live in space become available.

Early reasons for human space flight were socio-political. The Cold War pitted the Soviet Union against the United States. Competition between the two superpowers was the impetus for space flight success that has been a benefit to all. Scientific discovery is often a by-product of political or military actions. It was so in this case. National prestige of nations is built on an expanding commerce. The International Space Station prestige is still a factor in the human expansion into space. Economic growth and well-being of nations are built on an expanding commerce. The International Space Station (ISS) is an example of what can be accomplished through international cooperation. The United States and Russia are partners and are working closely together to build and operate the ISS.

When they realize the truth, a surprising fact to many is the realization that the American space program is a very minor part of the Federal Budget. Reporters and the mass media like to broadcast and publish the costs related to a space probe, a new telescope or human missions. Their approach often seems focused on dollars, not results. Space flight is one of the biggest bargains of federal spending. NASA's budget has been considerably less than 1% of the United States national budget since the end of the Apollo program. Less than one cent of every dollar that the government spends is used in the U.S. space program. This is a small price to pay for continuance of this source of new information and national pride that affects everyone every day. The space program has been a benefit to all of us.

The Apollo moon landings, Skylab, service and repair of the Hubble Space Telescope, Space Shuttles, and the International Space Station are all examples of the exceptional success of human space flight. In addition, there are all of the unmanned missions, telescopes, moon landers, Mars Landers, probes visiting other worlds, the list goes on and on. Collectively, they have provided a wealth of scientific information and technological innovations. Of even greater importance, they have been a stimulus for the youth of our country and the world to become interested in science and technology.

Change is in the offing. Until now, all space travel has been sponsored and paid for by national governments, but soon commercial space flight will begin in earnest. There have been a few tourist trips to the Russian Space Station Mir and the ISS, but the big change is the beginning of commercial, not governmental, space flight. The public at large has a deep and abiding interest in the exploration of the new frontier. It is estimated that 5,000 persons/year would be willing to spend $50,000 to travel to Low Earth Orbit (LEO). The number jumps to 30–40,000 passengers/year at a price of $25,000. The opening of space to the public only awaits a lowering of cost, a more economical access to this great adventure. Space travel will become a new economic driver in the future.

Scientific change and innovation due to the space program are a part of our everyday lives. The value to the American economy from space related spin-offs greatly exceeds the cost of the space program. In the future, beyond LEO, the moon will become a popular tourist attraction.

From a science perspective, astronomy research facilities on the far side of the moon will have the advantage of no atmospheric distortion and an absence of Earth-generated radio waves. On Earth, there are a number of radio-telescopes that are used to study faint radio emissions from the time of the Big Bang when the Universe came into being. Further, they search the sky for incoming signals that would come from intelligent life on other worlds There are two problems with radio telescopes on Earth. First is the immense quantity of radio and television signals that are continuously present in our atmosphere. Second, the ionosphere circling Earth that has charged particles. Both of these sources provide static that decreases the effectiveness of Earth based radio-telescopes. The moon, without an atmosphere and the ability on the far side for the moon's mass to block radio signal contamination from Earth means that the moon can become an astronomer's paradise.

One of the critical problems facing Earth's international economy and global prosperity in the coming years is available and affordable electricity. A Solar Power Satellite (SPS) was envisioned in 1968. Such a power station would be located far above Earth in geosynchronous orbit. Geosynchronous means that its rate of rotation around the Earth keeps it always above the same spot. The station would be in continuous sunlight, collecting solar power and transmitting it to Earth by microwave beams that can be converted to electricity. SPS technology is predicted to revolutionize electricity production without damage to the environment. Japan has indicated that it hopes to have a first SPS functioning by 2040. The problem with an SPS facility at this time is the cost of sending it into space. If there were a functional human colony on the moon, components for an SPS could be manufactured there and launched using electromagnetic propulsion for a fraction of the cost required to launch from Earth. A second possibility for launch of an SPS and many other items, as well as humans, from Earth, would be the success of the space elevator or some other advanced launch technology.

Our planetary neighbor, Mars, is another reason for us to travel to space. In the nineteenth century, better and better telescopes allowed observers to accurately examine Mars. Polar ice caps and "canals" were described and many felt that Mars was populated by an advanced civilization that used these canals to transport water for irrigation. All of those theories have now been disproved but many feel that Mars may have native life below the surface, where liquid water and warmer conditions are perhaps more suitable. If life is found on Mars, the similarities and differences from Earth life will have profound effects on our view of life's origins. That is the one reason for exploring Mars in person, the possible discovery of life native to that planet. We now know, based on information from the Phoenix Mars Lander in 2008, that there is water on Mars, and that it is near the surface. This is no longer supposition and wishful thinking. On Earth, water is essential to life, we assume that this will be true on other worlds. The ability to study completely new organisms and compare their innermost workings with Earth life would be a biologist's dream and broaden our perspective of what life means. When we begin to explore Mars in person, care must be taken to not introduce Earth microbes to Mars and then later "discover" that Earth and Mars have life with similar characteristics.

If it turns out that Mars is barren with no evidence of native life, another option is open. That is to terraform Mars so that it can become a second home for Earth life. Such a project is very long term but theoretically possible. The entire process would take over a century and cannot be seriously considered until humans have explored Mars and evaluated its resources. Terraforming includes two initial major projects; warming a planet to melt ice and creating an atmosphere acceptable to Earth life. According to proponents, both of these projects are feasible, but they must wait for Mars to be explored.

Terraforming Mars to become a second home for humans and other Earth life has several attractive aspects. It gives us room to expand. Humans are in surplus on our planet and increasing. Having a habitable Mars open to immigration would attract many adventurous settlers. From a long term perspective, in the past millions of years, Earth life has been devastated by major meteor impacts. This is likely to occur again at some time in the future. If Earth life, including humans, is present on another planet, we could potentially re-populate Earth following a cosmic disaster.

Part B: Star Bullets

Damn the torpedos, full speed ahead.

David G. Farragut (1801–1870)

There are serious dangers that long term voyagers or settlers on the moon or Mars must face. Stars, in addition to emitting light that bathes planets in their Solar System, emit a variety of particles and radiation sources. On Earth, we are protected from much of this radiation by our atmosphere and a magnetic field called magnetosphere. It extends far out beyond Low Earth Orbit. It blocks or deflects most incoming radiation. When we leave Earth to go to the moon or Mars or choose to live there, we will no longer be protected by our magnetosphere.

Increased radiation exposure during space flight is the greatest unsolved danger of long duration flight into deep space away from our planet and its magnetosphere's protection. In LEO, there is some increased exposure, but it is not a big problem for tourists or even those who spend a year or so. Most of the radiation is blocked or deflected around Earth. The danger is farther out on the moon or Mars or during travel to these destinations.

For those going to the moon or Mars, you need to know a bit more about radiation problems and some possible solutions. There are two sources of radiation. One is the Sun; the second is more random emissions from other stars of the Milky Way. The Sun, in addition to visible light, produces other emissions. For example, ultraviolet (UV) rays, although not seen, are short high energy rays present in sunlight. They cause tanning and sunburn to unprotected skin. UV light is easily blocked by a pane of glass or even sunscreen materials applied to the skin. Sunscreen doesn't really block UV, it absorbs it and converts it to heat that is then dissipated. The Sun also produces charged particles called protons. They are often referred to as ions to indicate that they have an electrical charge. Protons don't reach the Earth's surface but can affect astronauts in space, especially beyond LEO. Keep in mind, space flight from the standpoint of radiation is relatively safe until you leave our magnetosphere to go to another planet. If you plan to be just a LEO tourist, you can forget radiation dangers and let the high flyers worry.

Under most conditions, even the solar protons are not plentiful enough to be particularly dangerous to astronauts on the moon or on Mars. However, sunspots can explode, causing solar flares that are frequently associated with Coronal Mass Ejections (CME). CME hurl huge masses of material from the Sun into space and release great numbers of protons. These protons reach the Earth/moon vicinity quickly, as they travel at the speed of light, almost 11 millions/min. Explorers or settlers on the surface of the moon and Mars will need to be ready to seek shelter when a CME occurs, and minimize their exposure to the greatly increased level of radiation. At this time we are unable to predict CME's or solar flares. In January 2005, a great sunspot erupted hurling a billion tons of cloud of electrified gas into space. On Earth, we were protected by the atmosphere and the magnetosphere, and there was no increased danger here, but the moon and anything on its surface were exposed. There were not any humans on the moon at the time, so that this solar flare were exposed. Figure 3.2 is a photograph of a solar flare taken by the National Oceanic and Atmospheric Association (NOAA).

In the future, when the lunar outpost is a going concern or a lunar colony has been established, solar flares will constitute a real hazard. Fortunately, protons are easily stopped and are only a slight additional risk to lunar astronauts inside their space vehicle, lunar lander, or surface habitats. One plan is that habitats will be covered by a protective layer of lunar soil that

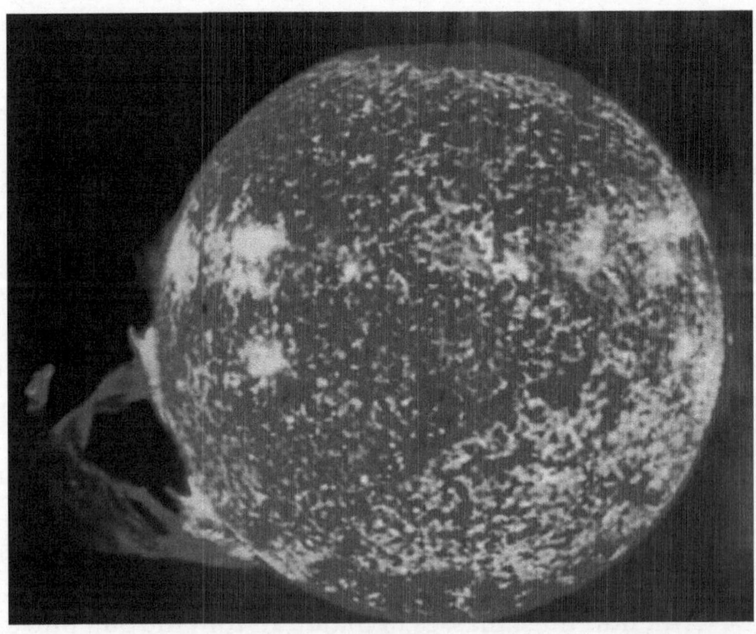

FIGURE 3.2 An image of a large solar flare on the Sun. (Picture courtesy of NOAA)

is called regolith. Regolith is the name for loose material, rocks, sand, and dust on the surface of bedrock. The danger develops when astronauts are out on the surface and may have only a suit for protection.

Space suits have some contradictory requirements. Above all, they need to be airtight to protect you from the vacuum. They also need to provide oxygen to breathe and a system for removing carbon dioxide. Finally, there is the matter of maneuverability. If you can't walk, bend, reach, grab, or twist, you can't get the job done when out on the surface of some new world. Spacesuits need to be flexible and easy to operate to permit exploration and to allow the lunar or Mars explorer to perform work. Because of that requirement, they are so thin that they provide only minimal protection from radiation. We have yet to develop a spacesuit that protects from radiation and still allows maneuverability. It seems you can't have both, at least not at this time. New technologies are often undreamed of till they happen. Never say never.

One proposed solution is to have a number of moon-base storm shelters or protective moon rovers nearby to be available to astronauts working outside when there is an unanticipated solar flare. Figure 3.3 illustrates the relative proton absorptive protection of a space suit compared to other shielding.

In addition to radiation from the Sun, there are other kinds of radiation that come through our Solar System from all directions. They are called Galactic Cosmic Rays or Galactic Cosmic Radiation (GCR). Most of these bad guys originate in the Milky Way.

Stars are like living systems on Earth. They have a finite life span. They can end their luminous existence with a mighty explosion called a supernova. When this occurs, immense quantities of radiation are released in all directions. Most of these cosmic rays are protons like the ones from the Sun and no more dangerous than the solar protons. They are easily stopped by the hull of space ships. A small portion of the cosmic rays are ionized nuclei of heavy metal atoms, particularly iron (Fe). These particles travel across the galaxy at light speed until they come in contact with an obstacle, such as a planet, a spacecraft, or humans inside the vehicle. They are hundreds of times more dangerous than the protons. They easily pass through spacecraft walls and through the

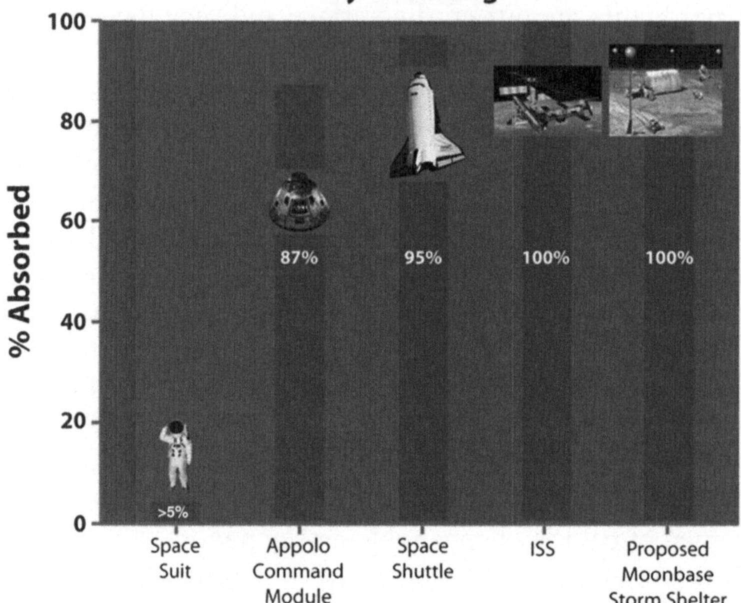

FIGURE 3.3 Solar protons are a danger when a solar flare erupts on the Sun. Astronauts with just thin space suits are not well protected. They will need rapid access to a storm shelter until the solar flare is over. This picture shows the relative ability of spacesuits and various spacecraft to stop solar protons.

astronauts. As they do this, their electrical charge ionizes everything that it contacts. The particles are magnitudes smaller than any conceivable bullet. They pass through the human body and we aren't even aware of it. The damage is caused by breaking chemical bonds inside the cells which disrupts normal cellular function. They can kill cells and damage DNA. Most parts of damaged cells can be repaired or healed by normal body mechanisms. The real danger is at the level of the nucleus within the cell. The nucleus contains our hereditary information or DNA. When that iron nucleus passes through the nucleus of cells, it can cut the DNA into fragments that may or may not heal correctly (Figure 3.4). When ionized cells with damaged DNA replicate or divide to form two daughter cells, they can pass on the wrong information.

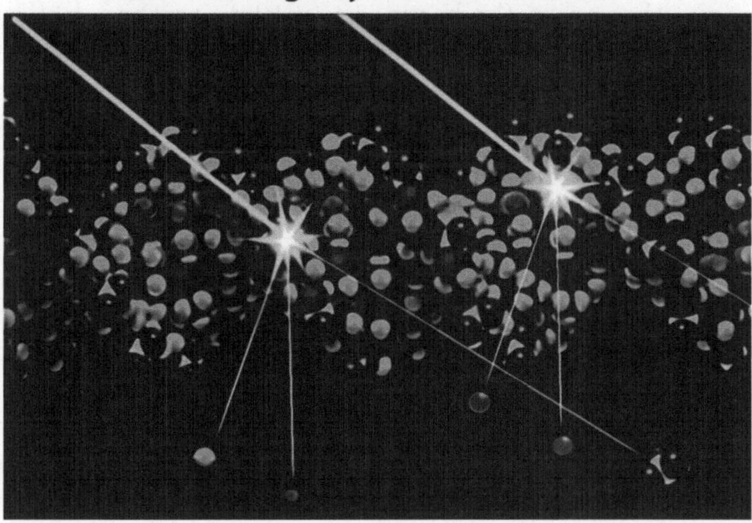

FIGURE 3.4 Galactic Cosmic Rays (GCR) which are largely high energy ionized iron nuclei, can cause permanent damage to DNA in the nucleus of the cells.

Such damage can result in an increased incidence of cancer. There can be damage to the DNA of male or female reproductive cells in the sperm or egg. Incorrect information at this level could result in gene mutations if the individuals were reproductively active after exposure. The damage caused by the GCR cannot be observed by the human eye, not even with a microscope. There is no mark indicating the ions entrance into or exit from the body or the body's cells, just as you cannot tell that you have been exposed to an x-ray.

To give an idea of the magnitude of radiation way from Earth without shielding, about 5,000 protons or heavy nuclei would strike your body every second. The protons would be stopped near the surface, but the heavy nuclei would pass completely through you and out the other side. There are no known instances of cancer occurring due to space flight away from Earth, but the number of astronauts who have left our biosphere and traveled to the moon is quite small and their journeys were relatively short. The only evidence that an astronaut has been struck by a GCR is when it passes through the retina of the eye. Apollo astronauts reported seeing flashes of light during their missions. This was due to ionization damage to the eye by the heavy iron nuclei. An abnormally high incidence of cataracts has been seen in the Apollo astronauts who went to the moon.

When we go back to the moon, astronauts will have to be protected by the vehicle that takes them there, as was done for the Apollo crews, or perhaps reside in a separate structure that serves as a habitat and laboratory. It may have been sent to the moon by remote command and be ready for occupancy when the explorers arrive. Accommodations at a first lunar outpost will be limited, but will grow in time to be a home away from home. After a moon base is established, the habitat will likely be underground or protected by covering exposed parts with a thick layer of lunar regolith. As an alternative, the moon, as well as Mars, has lava tubes left over from ancient volcanic eruptions. They are similar to lava tubes on Earth. They could be utilized as the outer framework of permanent facilities that would provide a thick layer of protection from GCR. Living underground in a tube or cave could also protect the astronauts and their outpost from micrometeorites and small meteors that frequently strike the moon.

Unfortunately, lava tubes may not be found in the spots where we want to develop settlements. Experience gained by learning to live and work in the moon's hostile environment will better prepare us to make the longer voyage to visit, and then establish, an outpost on Mars.

New lunar explorers will have longer tours of duty on the moon than the Apollo astronauts had. Exposure to GCR on the surface will be closely monitored and the total radiation dose will be limited to protect the astronauts. A record of total exposure will be kept and used as a basis for determining the length of residence on the moon. At this time, it appears likely that more permanent colonies will move deeper underground to limit radiation exposure. Increased radiation exposure will not be eliminated but kept within acceptable boundaries. With reasonable shielding, a few days or even several weeks of space radiation exposure is a relatively mild insult. A several year trip to Mars and back poses a serious risk of increasing and the likelihood of developing cancer and of damaging reproductive cells in the testes and ovaries that could cause mutations in offspring produced after the mission.

Based on our current knowledge, Mars voyagers will be exposed to a level of radiation that would not be acceptable for nuclear power plant workers or hospital x-ray technicians. One prediction by a NASA radiation expert in 2004 was that the added risk of cancer from a 1,000 day trip to Mars and back was somewhere between 1% and 19% for a healthy 40 year old male, and somewhat greater for a female because of the possibility of breast and ovarian cancers. This raises ethical considerations and the need for a thorough education and evaluation of the dangers for the individuals who volunteer to make such voyages. Some techniques being considered, or under development, will deflect or absorb the GCR before they have the opportunity to strike the crew during interplanetary flight. Humans will then be able to safely travel on long space voyages away from Earth.

Does this all mean we can't or shouldn't go to Mars or away from the Earth on other long voyages? No. Space agencies continue to seriously consider long lunar exploration missions despite these challenges. They are attempting to more accurately quantify the risk and to develop new methods of protection. It is possible to shield against GCR or deflect the particles with giant magnets. Weight constraints, due to the cost of space flight, prohibit such techniques at this time. From a radiation perspective, current technology is just not capable of producing a safe environment for traversing deep space on long voyages. In the future, new technologies may alleviate or perhaps eliminate radiation dangers away from Earth. For now, it is a problem, but exploring the Earth was, and exploring the ocean depths is, a problem today, but we have persevered and we will continue to do so. Nobody said it would be easy, or safe, to be an explorer or a member of the military during wartime, but many choose these vocations in part because of the risks. Damn the GCRs, full speed ahead. There's exploring to be done.

4. Change and Prosper

It is not the strongest of the species that survives, nor the most intelligent that survives, it is the one that is most adaptable to change.

Charles Darwin (1809–1882), Naturalist

A hallmark of life on our planet is the ability to change, to adapt. When we are exposed to new environments, we go with the flow. We modify the way our bodies function and become fit to live in the new environment. By so doing we become survivors and pass that ability on to future generations.

There are many diverse environments present on Earth. Plants and animals have had millions of years to develop traits that allow them to not just survive, but prosper, across an enormous range of environmental conditions. To live is to adapt. The ability to adapt is one of the great accomplishments or virtues of Earth-life, yet there is one environmental factor that has never changed on our planet. That is the gravitational force. All of the diverse environments on Earth still have that one thing in common. Gravity! It has been, and will be, constant as long as the mass of the Earth remains constant. The only way to get a decrease in gravity is to leave Earth and journey into space. Keep one fact in mind. Gravity is not just a good idea, "It's the Law."

Endotherms are animals that maintain a constant internal temperature by producing heat metabolically inside their bodies. Mammals and birds are the most successful endotherms, or warm blooded creatures. Warm-blooded mammals, including humans, are wonderfully diverse but internally very similar. Birds are the other group of endotherms but they are not quite as closely related to humans as our fellow mammals are. The distinctive feature about mammals and birds is that we continue to remain active regardless of environmental temperature. To accomplish this, we all adhere tightly to a principle called *homeostasis.* The simple definition

R.W. Phillips, *Grappling with Gravity: How Will Life Adapt to Living in Space?*, Astronomers' Universe, DOI 10.1007/978-1-4419-6899-9_4,

of homeostasis is "to remain constant." It refers to our internal environment, the substance of the body underneath the skin. In spite of a changing external environment, the body's control systems ensure that a mammal's innermost parts are protected from change. The internal body temperature operates within a narrow range that is maintained by elaborate and integrated control systems. Without these control systems, the rest of the animal world becomes lethargic or dormant in cold weather. They seek shade or go underground during hot times. The fact that homeostasis works so efficiently allows us to adopt an active life style that would not be possible in the environmental temperature of Earth.

How does homeostasis work? What and where are the controls? It really boils down to one word. The word is adaptation and it will be the byword of life's response to space travel. Adaptation is a basic attribute of all life, not just mammals and birds. It is defined as any change in structure, function, or behavior to make life more fit to survive in its environment. Without the ability to adapt, life would be confined to a few very small regions of our planet.

Life does not just inhabit, but prospers in every imaginable nook and cranny on Earth. Life is found wherever we look, desert to rain forest, mountain tops to the depths of the seas, tropics to the arctic, even deep under the surface in caves, rocks and soils. Life is everywhere. Small homeostatic/adaptive changes are continually being made by our body that allow us to unconsciously adjust to the varying conditions of our environment. That's the nice part of the system. You don't have to plan or schedule to adapt. Your body just does it. For example, your body temperature is constant regardless of temperature changes. Step outside from a warm room on a cool evening and your temperature control system will automatically activate. What happens then?

A series of homeostatic/adaptive controls maintain a constant internal body temperature in mammals and birds regardless of the environmental temperature. Overall control resides in a "thermostat" located in the hypothalamus, a structure deep inside the brain. It has two different areas that respond to either heat or cold. The temperature of circulating blood provides information on overall body temperature status. Blood flowing through the lungs and heart is distributed to all other parts of the body. The temperature of that blood when it reaches the brain can rapidly activate either the warming or cooling center. The homeostatic system is programmed to get rid of extra heat or conserve heat based on changes in blood temperature. There is a whole cascade of responses that come into play when the blood entering the brain is heated or cooled. When you move from a cold room to a hot room your body, including the blood, begins to get warmer and you need to dump that extra heat. One way is by radiation. Surface veins in the legs and arms dilate while deeper vessels constrict. Blood is pumping back to the heart just under the skin. The skin gets warm and heat can radiate away. Sweating or perspiration may also occur to maintain body temperature by evaporative cooling. So long as the humidity is not too high, sweat evaporates. Evaporative cooling is very effective in getting rid of excess heat.

When you move to a cold environment, the surface veins constrict. Blood flow in these veins decreases and blood is returned to the heart in deep veins next to the arteries carrying blood out from the heart to the limbs. With the two vessels next to each other and blood flowing in opposite directions, a countercurrent heat exchange is set up. As the temperature in the exposed extremities drops, blood is cooled and returned toward the heart. Running next to the warm arterial blood leaving the center of the body, the cool venous blood is warmed and the warm arterial blood is cooled. With this intricate but functional system operating, deep body temperature is maintained and skin and extremity temperature will decrease.

We have preserved and continue to use goose bumps, an ancient warming technique that doesn't provide benefit any more. Goose bumps are nothing more than contractions of tiny muscles around follicles in the skin that would cause hairs to become erect if we had a fur coat. Our distant ancestors presumably had a thick pelt and the erection of the hairs would have increased body insulation for them. Another internal warming technique animals do is shiver. Shivering is a non-coordinated contraction of opposing muscles; extensors and flexors. No useful work is done, but the muscles are active and generate heat as they contract, helping to keep the internal body temperature constant. Shivering is not an effective long-term solution to maintain body temperature, but it works well for brief periods.

Increasing metabolic rate by the action of thyroid hormones is effective in the long haul. Increased thyroid hormone synthesis and secretion is important during long term exposure to cold. Huddling to share body warmth or curling up into a ball to decrease exposed surface area and heat loss are other common behavioral activities in animals needing to conserve heat. The fact that so many processes come into play to maintain internal temperature indicates how important that regulation is to our well-being.

Adaptation is part of the story, but there is more. In the words of Paul Harvey, noted radio commentator; "Here is the "Rest of the Story." Adaptation is coupled with evolution, the other response to environmental change. Traits brought about by random mutations that are beneficial are continued in our DNA. They provide that little edge that allows passage of a new advantage to future generations. From the standpoint of temperature control, animals frequently exposed to cold often develop a thick pad of fat under their skin that serves as insulation. This occurs in seals, dolphins, whales and also penguins and polar bears. Many animals develop a thick winter coat to conserve heat, then shed it in the warmer months. Although many of these changes are adaptive, they are also evolutionary. Beneficial traits have ultimately become a part of species or individuals' heredity.

In some cases, communities of animals have unexpressed or perhaps unused genes, waiting to be used when the need arises. These traits can be structural modifications and appear or disappear depending upon the environment. The changes appear to be evolutionary, but are really the expression of latent genes that are rapidly expressed as needed during environmental fluctuations. This does not mean that evolution did not occur in the past. Instead, life on Earth conserves genes that allow them to rapidly develop different form and function to deal with changing environments. Some environmental changes come and go. Being ready to change/adapt during droughts as well as plentiful moisture ensures survival.

An example of this very thing is demonstrated by the finches in the Galapagos Islands that helped lead Charles Darwin to his development of the Theory of Evolution. They rapidly change form as their environment changes. When the islands receive plentiful rains over several years, food is abundant and the shapes of the finches' beaks are similar, and with lots to eat, the finch population expands. During periods of drought, when food is scarce, the population decreases and the beaks of the various finch species become distinctively different. This beak specialization allows each species to gain nourishment from the different plants that grow under marginally arid conditions. Even without weather charts from that period of time, we assume that Darwin studied the Galapagos finches during a period of drought. Under those conditions the beaks of the finch species would have been different. If he had been there when rain was plentiful, he might have missed this opportunity to further develop his theory. The capability to turn on or off certain genes to adapt to a changing environment is a clear-cut example of life's ability to respond to changing conditions.

Humans can also change. In a modern society, we carry our environmental control technologies with us and may not become exposed to new environments. By using furnaces and air conditioners, it ensures that we live in a controlled environment. We will eventually lose our ancestors' ability to easily adapt to natural fluctuations in the climate.

A change in gravity is a unique experience that life on Earth had never encountered until the space age began. Sputnik 2 was the second space launch that ever occurred and it carried the little Russian dog, Laika, into Low Earth Orbit. She was the first creature who ever experienced microgravity. Before her trip, there were concerns that Earth life, including humans, might not be able to survive in space, even within a vehicle. It was a short trip and Laika did not have an opportunity to adapt to her strange new environment, but she proved that animals could live in space (Figure 4.1). That was a big step and opened the door for the rest of us.

With the advent of space flight, we have opened a new chapter for Earth-life. To succeed as we move out from our home planet, we must learn to adapt to new gravitational fields. That will be a new experience; something that has never been done before. In the near future, we'll inhabit, in a more or less permanent fashion, spacecraft and distant planetary surfaces. For now, let's consider space as any new and unique gravitational environment away from planet Earth. Some new environments may be space itself, where there is no effective gravitational field and everything loose appears to float while in free fall. Other environments might include the surface of the moon where gravity is 0.16 of the Earth's gravity or the surface of Mars with a gravitational force of about 0.38 of that on Earth. It may even be possible to occupy a large asteroid that would have

FIGURE 4.1 The small dog Laika was the first animal, really the first Earth-life, to go into space and orbit the Earth. She is shown in this picture with the Russian engineer and designer, Sergei Korolev, who was a major force in developing the Soviet space program. Laika was launched on October 4, 1957, on Sputnik 2. Heart rate, breathing rate, and blood pressure were monitored.

only a very slight gravitational field. Wherever it is, you will have to adapt in order to prosper in your new environment.

A potential source of gravity while in space is centrifugation. That's the dream of many space venture pioneers who are eager to permanently move away from Earth. They envision a torus, a large doughnut shaped orbital space habitat that spins, creating a gravitational force that is greatest at the inner surface of the outside rim. At this time, no serious efforts or funds are devoted to developing such large and complex space worlds. That will come after we have settled existing worlds like the moon or Mars.

As we move out and begin to develop new outposts or colonies in our Solar System, our bodies will change. Gravity has shaped us to live on Earth, not in μg or in lesser or greater gravities. Unique changes in structure, function, and behavior will occur. These changes will make us more fit to live in these new and unfamiliar gravitational fields. As we become more fit to live in the new gravitational field, we will become less fit to live on Earth (Figure 4.2).

You cannot chose to adapt or not. Adaptations have never been under conscious control. While living on Earth, a person,

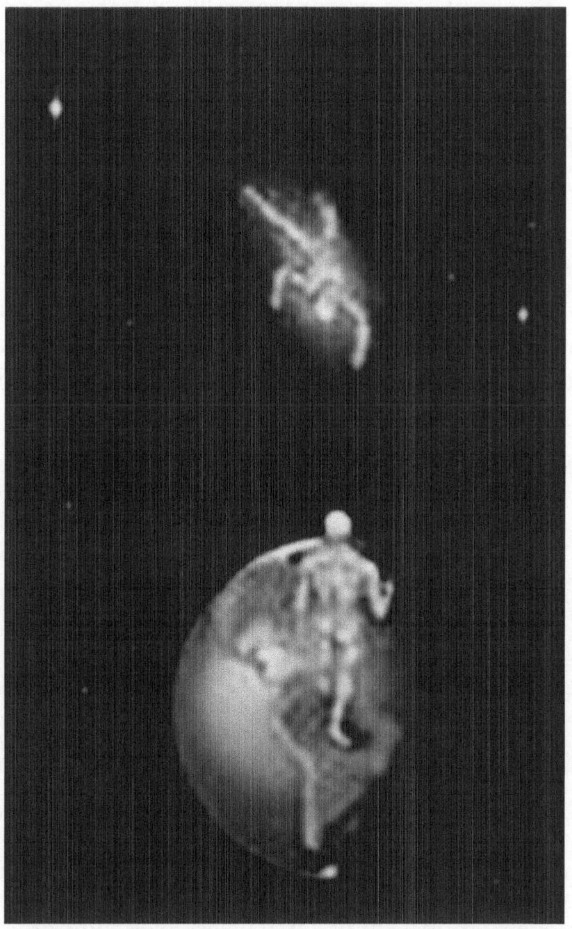

FIGURE 4.2 At any given time, a person can be maximally fit to be on Earth or in space, but not at the same time. The environments are sufficiently different that body structure and function must change as we move from one location to the other.

other animal, or plant is maximally fit to be on Earth and in its gravitational field, but less fit to be in space. Because life is so malleable, it begins to change as soon as the μg environment of space flight is reached. Over the hundreds of millions of years that life has existed on our planet, the variety of life forms that we recognize today or that have been present in the past have developed an amazing ability to change form, function, and behavior to cope with the changing conditions that our planet has to offer. Plants and animals that have evolved on land are able to challenge gravity most directly by developing structures like trunks and stems or legs with hollow bones that allow them to defy gravity.

There are countless examples of how various animals have adapted to space flight. The comb that honey bees build, when they first enter space, is not as uniform or structured as those produced on Earth. As times in an μg environment goes on, the space comb that is built is regular and uniform as on Earth.

Two spiders were carried to Skylab, the United States first space station, that was launched May 14, 1973. The spiders changed after they arrived at the lab. The initial web that they built did not have the symmetry we are used to seeing in webs that are constructed on Earth. After adaptation, a new web was constructed that was similar to ground-based webs (Figure 4.3). What is more

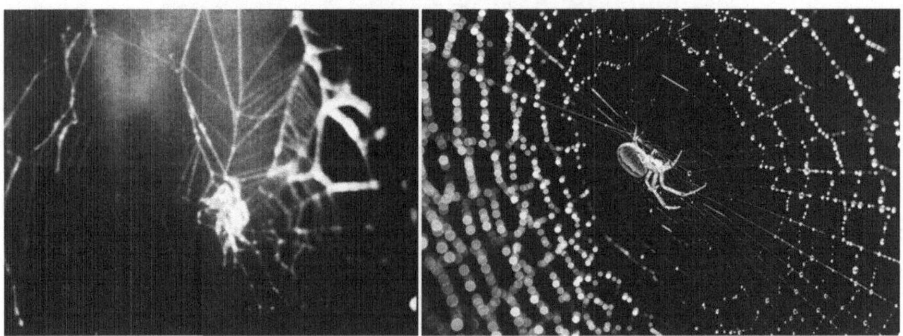

FIGURE 4.3 During the Skylab missions, two spiders were taken to the lab. Shortly after arrival, *left frame*, the web constructed by one of the spiders was haphazard without the symmetry usually seen in spider webs. After adaptation to its new environment, *right frame*, the spider constructed a more uniform web reminiscent of Earth webs. Even more amazing, the web fibers were smaller in diameter with a lower tensile strength. The spider sensed that less strength was needed to support itself in a μg environment.

astounding, the web fibers, formed after space adaptation, had decreased diameters and tensile strength compared to similar webs formed on Earth. The spider's control systems, at some level, were aware that in the free fall floating environment of space it did not require as robust a web to support itself. The reverse is also true. Spiders on Earth, that have their body weight increased, build a more substantial web.

When rats were brought on board a space shuttle in order to study adaptive similarities between rats and humans, a delightful but unanticipated behavioral adaptation was seen by the crew. The water supply to one rat's cage ceased to function and the rat needed to be given a drink. The crew moved the rat from its cage to an enclosed work station. A syringe, without a needle, was filled with water. A ball of water was expressed from the tip forming a sphere. The rat, seeing the water, reached out with its forepaws, grasped the ball of water and carried it to its mouth. It drank while holding the sphere in a way that cannot be accomplished in Earth's gravitational field (Figure 4.4). This was a truly unpredictable and effective act by this rat that could not have been learned or even simulated on Earth. It had adapted very well to its μg environment.

At some unknown time after entering space, body systems will become maximally fit to be in μg and less fit to function on Earth (Figure 4.5). The time required for this to occur varies. Fluid changes are very rapid, while bone restructuring may take years to complete. In all of your body systems, it will be a constant and continuing process, 24/7.

Our wonderfully adaptable body systems have allowed Earth life to comfortably live almost anywhere on Earth. This adaptation to space presents a problem to all space agencies with a human space program, or an interest in future human space flight. The goals of these agencies are to prevent or counteract space adaptations so that crew members can easily return to Earth after spending significant time in space. To achieve that goal, they are continually developing procedures called countermeasures. Countermeasures are designed to block or reverse space adaptation during space exposure. Most countermeasures are centered on fluids, bone, muscle, and cardiovascular systems. Such procedures are only minimally effective or even non-effective in some cases. But why?

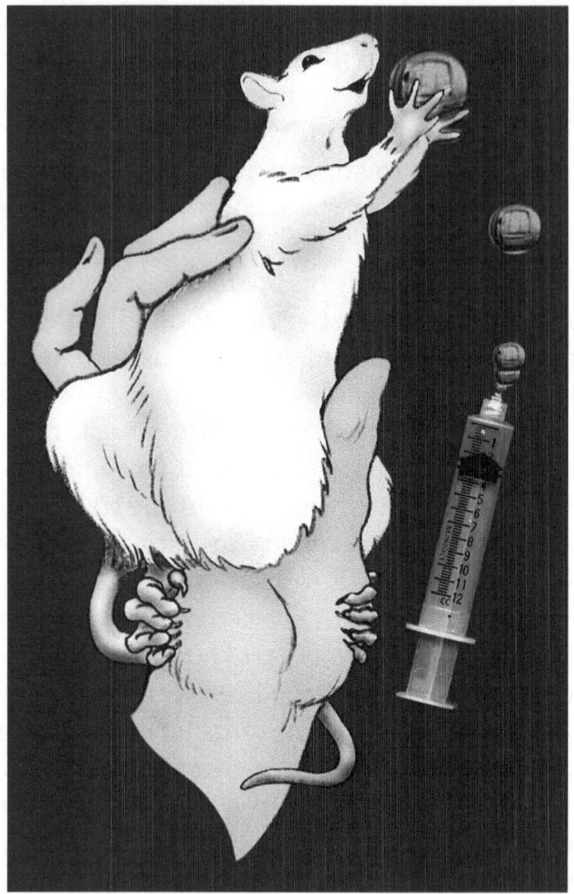

FIGURE 4.4 While in space with a broken water supply device, it became necessary to supply a rat with water in another way. The rat was removed from its cage and a large drop of water was expressed from a syringe by one of the astronauts. The water, in μg free fall, formed a sphere. The rat grasped the water in its forepaws, brought the sphere to its mouth, and drank. He repeated the process with a second sphere. Such an action is impossible on Earth.

There are two opposing actions. One is Mother Nature dictating that adaptation must occur to increase fitness in the new environment of space. The second is the countermeasure program developed to block adaptation, and permit spacefarers an easy return to Earth. Adaptations begin as soon as you reach space and are exposed to μg. They continue until adaptation is complete. The important thing to remember is that adaptation will be a continuous process

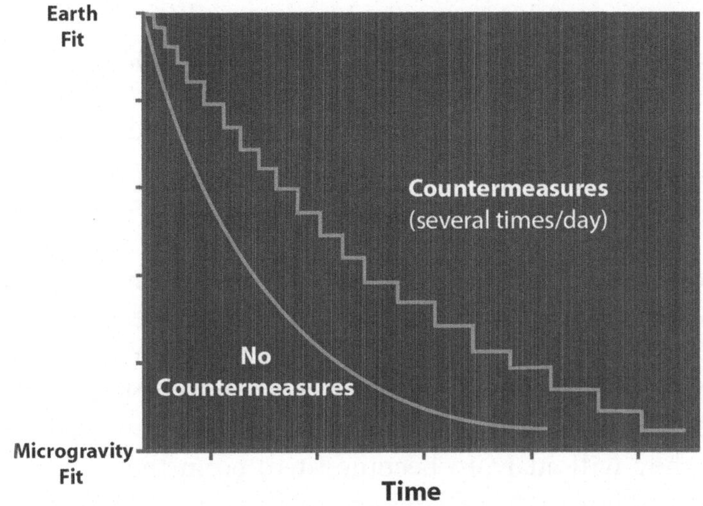

FIGURE 4.5 Adaptation to the new μg environment will continue until a new stable "adapted" point is reached. Countermeasures may slow the rate of adaptation but the end result will be the same unless the countermeasure is in place 24 h a day, 7 days a week. This is not a practical solution.

as long as you remain in μg or in a decreased gravitational field on the moon or Mars.

Popular countermeasures have been used include a variety of exercises designed to strengthen or stress muscles, bone, and the cardiovascular system. Others are intended to restore normal fluid distribution in the body. Undoubtedly these procedures have a benefit on a short term basis, but your body's processes of adaptation are continuous. On a generalized time scale, the adaptive processes and effects of countermeasures might look as shown in Figure 4.5. Countermeasures may slow, but they will not reverse adaptation brought on by exposure to μg. The final end point will be the same with or without countermeasures. The reason for this is that countermeasures are periodic and adaptation is continuous. You cannot exercise for 24 hours a day. During the time that you are not exercising, you will be adapting. The time scale is not specific, as the different body systems change at different rates. Fluids and muscles adapt rapidly while bones are slower. An end point will ultimately be reached as long as you stay in the new environment.

It is not that countermeasures are of no benefit, but they are only partially effective and may have no noticeable effect in the long term, after perhaps years of living in a different gravitational field. Long term residence on a planet with a greater or lesser gravity field than Earth will cause adaptation to that planet's environment. Your body systems will change and become appropriate for your new home.

The bottom line seems to be that designers of these many countermeasures don't seem to fully appreciate that "you can't fool Mother Nature." This is a fact we must learn to live with as human space journeys become longer. Countermeasures can be of benefit in short term flights, making it easier for space travelers to return home. As humans reach out to explore other domains of our Solar System, with new and different gravitational environments, they will adapt to become fit to be in that environment. This adaptation may well lessen their fitness to return to and live on Earth. Earth born creatures, including humans, who have adapted to a decreased gravity field, will be able to re-adapt upon return to Earth, but this is likely to be a slow process. You may spend months or years becoming re-adapted to your native gravitational environment.

Adaptation to a new environment should not be a basis for staying home. Instead, you will need to be prepared for the inevitable. You will change to be better prepared to function in your new quarters. Your body will take care of ensuring that you can continue to function; even if it seems a bit strange at first. Relax and let it happen, that is a lot better than trying to fight the system.

5. Everything Is Topsy Turvy

Microgravity so free
Yet structured, so differently
What an adventurous place to be!

Logan Mather (1968–Present)

Part A

What happens to gravity? Is it really zero when we go into space? After all, everything that is not fastened down floats, including astronauts and their belongings. Surprisingly, zero gravity does not actually exist. All solar systems in the universe, stars and their orbiting planets are held together by gravity. Earth's gravitational field holds the moon in an orbit about 238,000 miles away. The moon's gravitational field causes the tides in Earth's oceans. All of the planets and moons in our Solar System are held in stable orbits by the Sun's gravitational field. Zero gravity is not an accurate description of space flight conditions.

As a further note the International Space Station has enough mass that it creates a gravitational field. Experiments on the station that require the least gravity possible in order to succeed, are physically located near the center of the station's mass. In that small area the gravitational field is negligible. It is often called the 'sweet spot' by investigators who desire the least gravitational disruption of their experiments.

What is gravity? According to Sir Isaac Newton, gravity is the force of attraction between two masses. We are attracted toward the center of Earth and Earth is immeasurably attracted toward our center of mass. The term microgravity, abbreviated μg, is often used to describe the environment you encounter and live with in space. Gravity is there but our bodies can't detect it. Microgravity is a reasonable description of what we are exposed to during space flight.

R.W. Phillips, *Grappling with Gravity: How Will Life Adapt to Living in Space?*, Astronomers' Universe, DOI 10.1007/978-1-4419-6899-9_5,

Gravity is also a force of acceleration. Near the Earth's surface it is 9.8 meters per second, per second. When falling, you travel 9.8 m the first second and 19.6 m the next second and so on until you reach terminal velocity. Terminal velocity is due to the drag of the atmosphere. It works to slow you down during a long fall as you pick up speed. The concept of gravity's acceleration is simple (Figure 5.1). Even young children realize that the higher the jump or farther the fall, the harder the landing. They just don't call it acceleration.

Actually, Earth's gravitational strength in LEO is about 10% less as when you are standing on the surface. The mathematical method for determining the gravitational force while in space is shown in Appendix 2. That still leaves the question, if it is not zero gravity in space, why do crew members and their articles "float"? The answer is simple. The spaceship in LEO is in free fall back to Earth. At the same time it is moving in a straight line and would leave the Earth's vicinity if it weren't for gravity. Articles

FIGURE 5.1 At an early age, we learn that gravity is a force of acceleration, even though we don't call it that. Small children know it is safe to jump to the ground from a chair, a bit more risky from a stepladder, and dangerous from a roof of a building. Every second that you are in free fall, your speed increases by 9.8 m until air resistance slows, and then stops your acceleration. That speed is called terminal velocity.

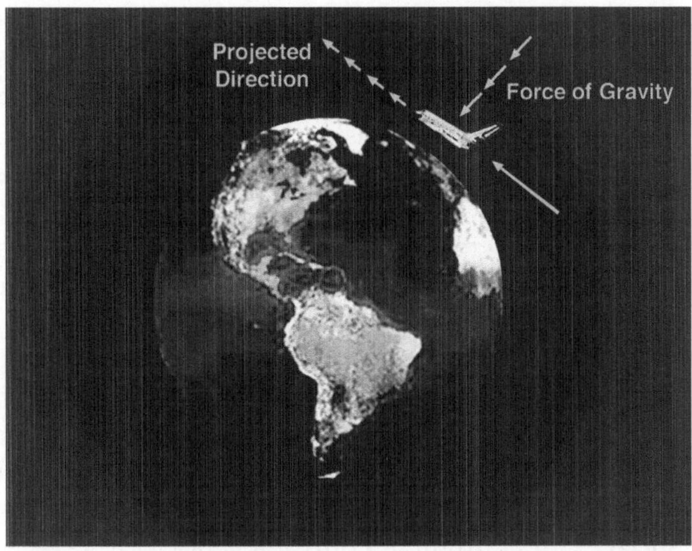

FIGURE 5.2 The spacecraft is traveling in a straight line, but is also in free fall towards the center of the Earth.

and people within the spaceship are floating because they all are in free fall together toward the center of the Earth (Figure 5.2).

It is the space vehicle's forward motion, which is 17,500 miles per hour in a straight line several hundred miles above Earth's surface and its rate of free fall towards the center of the Earth that keep the vehicle in orbit. The spaceship falls about 15 ft/s and the surface of Earth gets about 15 ft/s further away due to Earth's curvature. The vehicle is in an orbit that is circular or nearly so. The astronauts within the spacecraft have no sense of falling because there is no visual reference to establish the fall. This can be compared to the rapid descent of an elevator in a tall building. When the elevator starts down, you initially feel lighter but when the downward acceleration stops and a constant speed is maintained, you can't tell that you are in a controlled downward motion or even that you are moving.

Living inside a space vehicle or upon another planet for years, decades, or life spans will be a new experience for us. Before space travel came along, no one nor anything living on our planet had ever been exposed to a decreased gravitational field. Skydive from an airplane and you can experience free fall in the open, but you know you are falling and it is only a brief experience. It's not the same as LEO and free fall in a space vehicle. In a space vehicle, there is no sensation of falling, only a unique new freedom. You don't walk, run, or jump; just push off one surface and float across the vehicle to the next wall. This is not floating in the sense that you are supported by water in a swimming pool. But the word "floating" pretty well represents free-fall in space. It is the best description of life in orbit that we have. Continuous floating like this is a new experience for Earth life.

Floating in space is often described as being weightless. Technically, this is not right either as you would not be in free fall if you were truly weightless. However, orbital space flight produces the feeling of being weightless. During free fall on a spaceship, you don't recognize up or down without a visual cue such as lights, writing surfaces, or equipment control levers.

When you are in Low Earth Orbit and want to return to Earth, the solution is simple. **Slow down**. Using the space shuttle as an example, here is what you need to do. First turn the shuttle around until it is facing backwards. Briefly fire the engines. The vehicle slows enough that falling is faster than the movement forward. The spacecraft slowly sinks into the atmosphere where atmospheric drag slows it further. It circles back to land on Earth. In reality, the timing of firing the engines and the strength and duration of backwards thrust are critical. This is necessary for the descending shuttle to reach its landing field. With the shuttle going 292 miles/min, there is not room for error.

Drifting effortlessly inside a falling spaceship is a wonderful experience compared to other aspects of space. The space environment is very unfriendly. It is an almost complete vacuum that would cause you to essentially explode if exposed without the protective atmosphere inside a space suit or spaceship. Actually, you wouldn't explode, but all of your body fluids would begin to boil due to the lack of atmospheric pressure, and you are mostly made of water. This would be a rapid and unpleasant end. To be outside in space, on the moon or on Mars, a spacesuit is a necessity, not a luxury.

In addition to free fall, extreme temperatures, and vacuum, there are other major changes to get used to when you go into space. Space voyagers must rethink their perception of how things work.

Consider water or fluids in general. Everyone is familiar with a drop of water and how it flattens out when placed on a surface, yet its molecules remain intact. It stays together because of surface tension. Surface tension is a prominent factor in fluid behavior in μg. In liquids, individual molecules have a strong attraction for each other. These adherent molecules on the surface of a liquid act as a thin membrane stretched over that surface. Consider the bubbles that children create by blowing on a film of soap solution in a plastic wand. Surface tension creates these spheres from a thin film of soapy water surrounding a small amount of air. The reason that they are round is that a sphere has the smallest surface area of all possible configurations. The tension of the film pulls it into the spherical shape. In microgravity, all free fluids will be spheres when floating. Even when resting on a surface, a drop of water is spherical (Figure 5.3). There is a bubble of air in the sphere. This bubble does not rise to the top as it would in a functional gravitational field.

Another effect of μg on fluids is that hydrostatic forces no longer operate. Water does not run downhill. You cannot empty a cup or glass by turning it upside down. Water drops left or placed in the air will remain there. Some caution is necessary as free water cannot be left to float around in a space vehicle. It could be inhaled or cause damage to sensitive equipment. Hand washing facilities,

Figure 5.3 A sphere of water with a bubble of air inside rests as a nearly perfect ball on a leaf while in μg.

FIGURE 5.4 The hand washing device in the space shuttle galley.

such as the one in the space shuttle galley, have been designed to contain water. You place your hands inside the hand washing station and activate a water spray with your feet. Water is contained by air flow pulled into the washing chamber, which also helps dry the hands (Figure 5.4).

Refrigerators and freezers depend on convection currents to distribute the colder air generated by coils. As air is cooled, it becomes denser, sinks, and creates the currents. Conversely, air rises as it is heated and expands. That is why cooling or freezing coils are only needed in part of the refrigerator or freezer, yet keep the whole appliance cold. A freezer in space will not work that way because heat doesn't rise and cold doesn't sink. Instead, mixing of the air is dependent upon a fan. The fan is needed to ensure that cold air distribution is uniform.

Flames in space behave in a different manner than they do on Earth. In space, gases expanded by the heat of the flame don't

FIGURE 5.5 On the *left* is a match burning on Earth. Photo on the *right*, a flame in space.

rise (Figure 5.5). The flame of a candle in space is small, spherical, and cooler in temperature. Without air movement, all the oxygen near the flame is consumed and the candle will slowly become extinguished.

Microgravity and the feeling of weightlessness are what make space flight such a unique adventure. There is no way to prepare for such a major environmental change. You just have to experience it in person!

Part B: Let's Take That First Trip into Space

STAIRS

Here's to the man who invented stairs
And taught our feet to soar.
He was the first who ever burst
Into a second floor.
The world would be downstairs today
Had he not found the key;
So let his name go down to fame,
Whatever it may be.

Oliver Hereford (1927–Present), Poet

After launch the space vehicle arrives in Low Earth Orbit (LEO) and starts to circle Earth. Those onboard are in a free fall floating about the spacecraft. Up and down have no meaning except when you gaze out a window to look at Earth. There is a new order of daily living. Traveling at 17,500 mph, you circle Earth every 90 min. Sunrises and sunsets come in rapid succession. To see that many sunrises and sunsets can be disturbing to our internal daily or circadian rhythm. Since life began to evolve on our planet, the daily 24 h rhythm of dark and light has been with us. It is not easy to give up or ignore such an inherent rhythm. Many travelers in space avoid adapting to this rapid rhythm of light and dark by still maintaining a 24 h day. They continue to have three meals and only one scheduled sleep period in spite of the frequent light and dark cycle occurring outside the vehicle. On the space shuttle it also helps that there are a minimum number of windows as the work schedule leaves little time to spend gazing at the Earth.

The schedule will change on future tourist trips to LEO. Earth viewing will be the principle avocation. It is a never ending panorama of fresh scenes. The reason for this is simple (Figure 5.6). The space vehicle always travels over the center of the Earth in a straight line, yet its track is tilted so that it goes the same distance north and south of the equator. When launching from Kennedy Space Center in Florida, the vehicle must have at least a 28° tilt, as that is the launch site's latitude north of the equator. While the space vehicle travels in a straight line with a constant tilt, the planet is not stationary. It continues its rotation, one revolution per 24 hours. Each time your space ship circles the Earth it is approximately 22.5° west of the previous orbit because the Earth has rotated 22.5°.

The space vehicle will supply you with an Earth-like atmosphere. It must remain pressurized, supply oxygen, get rid of carbon dioxide as well as collect and contain liquid and solid wastes. On short missions, for example a couple of weeks, body wastes are stored and disposed of when you return to Earth. On longer missions, recycling water as well as oxygen is necessary.

Eating is a requirement for continued health and well-being. This will not be a problem on short tourist trips in LEO, just up for a few days to observe the Earth and enjoy space flight. You won't make the trip because of the cuisine. Simple meals that

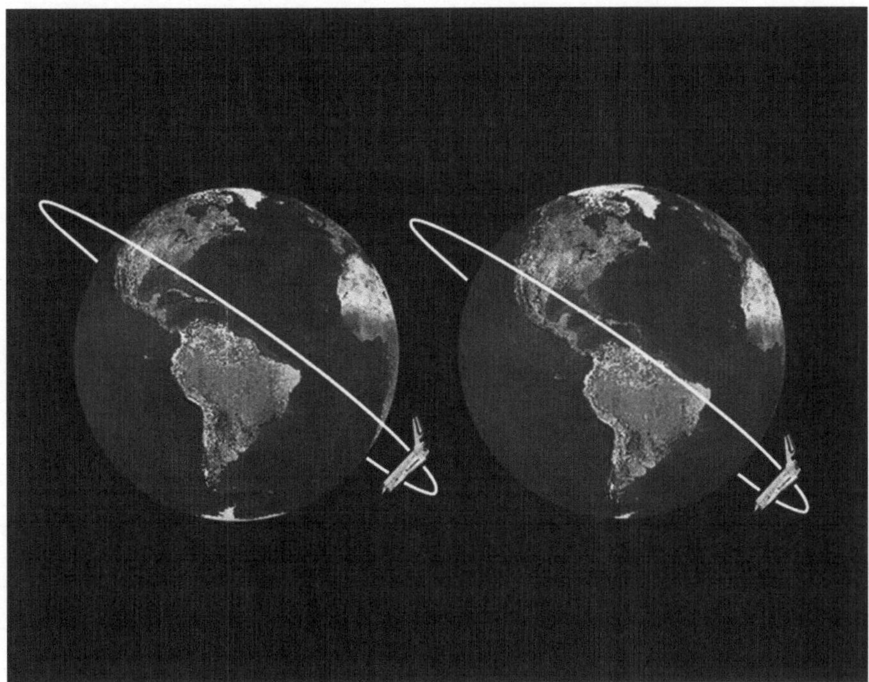

FIGURE 5.6 In this picture the Earth turns to the east or right. The space vehicle travels in a straight line at an inclination above and below the equator, while the Earth turns below it. With each circuit of the Earth, the vehicle is approximately 22.5° west of the previous orbit. Space travelers have a continually changing view of the Earth's surface.

require little or no preparation will be just fine. An adequate food supply and proper nutrition become more important on longer missions. From a nutritional perspective, it is a relatively trivial matter to go on even a 2-week LEO mission on the space shuttle. The shuttle carries plenty of food with quite a bit of variety. The nutritional challenge on the space shuttle has been to get the crew to eat enough and to not lose weight. There are several reasons for this. Early mission space motion sickness reduces some of the crew's interest in food. Just like sea sickness or motion sickness in a car, you aren't ready to eat a hearty meal or perhaps any food at all. Also, the astronauts on the shuttle have very tightly scheduled days. This will not be a problem for tourist type missions, but it has caused concern on some shuttle flights. For the crew, the desire to finish assigned tasks takes precedence over eating.

Mission success is a strong driving force. No one wants to return to Earth with the job not completed. The desire to succeed overrides the minor discomfort of a skipped meal.

Although providing an adequate diet for a couple of weeks is not a serious problem, a 3–6 month stay on a space station or an extended journey to a moon base or to Mars requires more planning. Not only is it necessary to supply the actual nutrients required for health and well-being, but for optimal performance on long missions, crews will need to have appetizing meals available. In order to encourage eating, space crews need to be able to look forward to an enticing and varied diet like they eat at home. There is another factor to consider in developing diets for long space flights. Today on the ISS and in the future on exploration missions, there will be international crews. Dietary preferences between cultures are often quite different. Meals will need to be appetizing to all.

Today we know that there aren't any major differences in nutritional or dietary needs between being in space and being here on the ground. Space foods that are commonly used today are similar to the kinds of food that you might take on a camping trip. Compared to preflight diets, they are lower in fats and higher in carbohydrates. Protein content of space diets is similar to the protein content of Earth diets. Most diets are low in fiber and, as on Earth, this can cause constipation. Part of the basis for using low fiber is the relative inconvenience of space toilets at this time. New technological developments in microgravity toilet design may solve that problem, but it is not available yet.

In today's modern space flight, the kitchen concerns are what foods to serve, crew preferences, and how to increase nutritional intake. With technological advances in the industry of food technology it is easy to forgot how wrong early suppositions were regarding the physical act of eating and drinking in space. Some were concerned that astronauts would not be able to swallow when they got to space. They feared food and water would not flow down the esophagus without gravity's help. It would have been easy to show how wrong they were by enlisting the aid of helpful friend to stand on his or her head and drink a glass of water through a straw. The water easily flows against gravity to the stomach by the peristaltic contractions of the esophagus. The act of swallowing in space is no different than on Earth. Those particular concerns were soon resolved during the initial Mercury missions and the early Soviet space program.

On short term shuttle missions or to resupply the ISS, food is prepared on Earth and shipped in a form that is ready to eat or requires little preparation. This approach will also be used on early moon missions and may also be used on the first Mars' missions. The final goal will be food production facilities at moon bases and on Mars when permanent outposts are established.

Journeying to space today, one of the problems is limited refrigeration for food. Food and drink can be heated or eaten at room temperature. Imagine in today's culture spending several months dependent on prepackaged and dehydrated food. It would be like an extended camping trip as noted in Figure 2.1. In essence, this describes space travel today. We carry all necessary supplies when we leave here and return when they are expended. New supplies are sent to space stations as needed to feed the crews. NASA dieticians and nutritionists strive to provide well balanced meals that will supply all nutritional requirements and be readily consumed. Therein lies a problem. On long missions like Skylab, Mir, and the ISS, the general behavior has been to consume favorite foods first, as long as they last. Attention to daily dietary requirements is somewhat disregarded. Crews do receive the necessary nutrients, but not always on the schedule that dieticians would prefer. Really, not too different from our behavior here on Earth.

Foods for consumption in space need to be low in mass because of the current high cost of sending materials into space. Some examples of currently used foods are shown in Figure 5.7. NASA utilizes, when possible, off the shelf products. In space, flour tortillas have largely replaced bread. This is due to their lack of crumbs, which are difficult to control in microgravity (Figure 5.8). A favorite food on the space shuttle is freeze dried shrimp cocktail. It is quite good unless you try to eat it immediately after re-hydrating. It can be a bit crunchy. If you wait till the water is all absorbed, it is a treat.

Eating and drinking in space is a bit different than we are used to on Earth, but it is not a problem to learn the new system. Then there is the rest of the story, what comes in must go out. Voiding body wastes while floating in microgravity is another "new experience." The system used on the space shuttle is designed on an

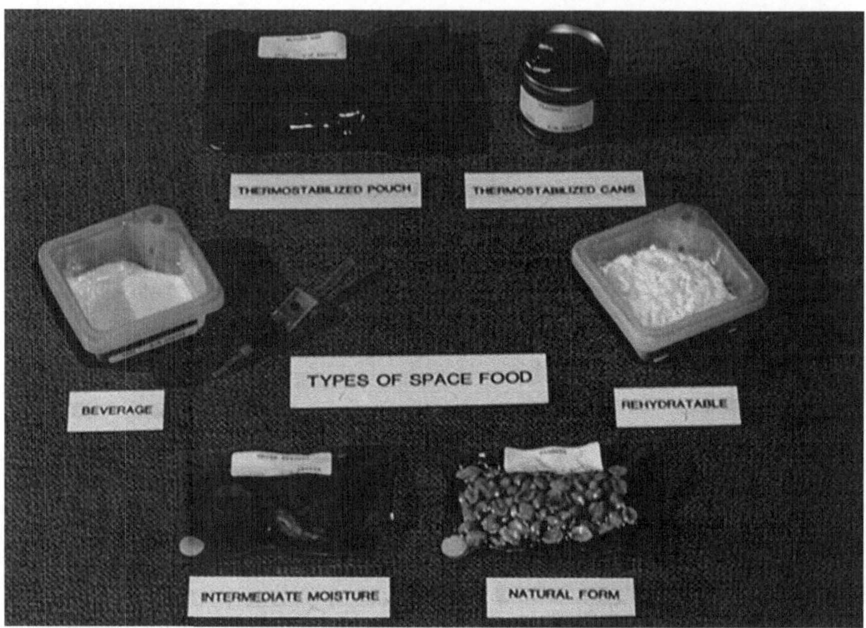

FIGURE 5.7 Various foods used on the shuttle and on the ISS; slices of beef with barbecue sauce, a can of fruit, dehydrated eggs, dried apricots, peanuts, and a drink. (Photo courtesy of NASA)

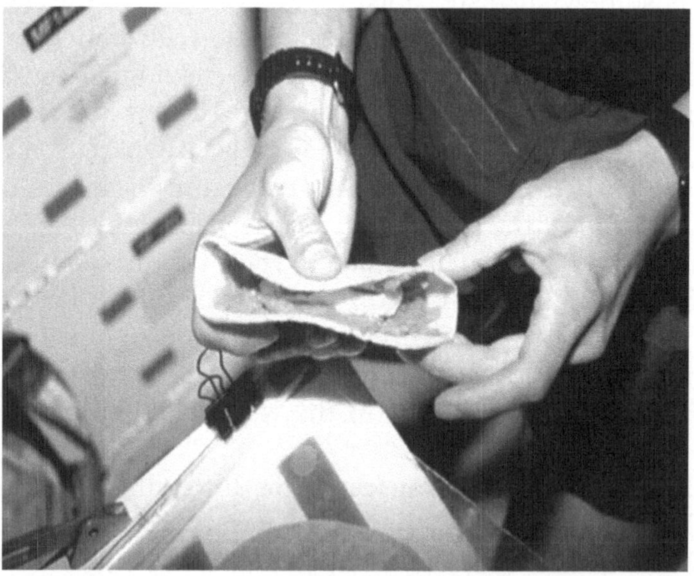

FIGURE 5.8 Preparing a lunch from a flour tortilla and refried beans. (Photo courtesy of NASA)

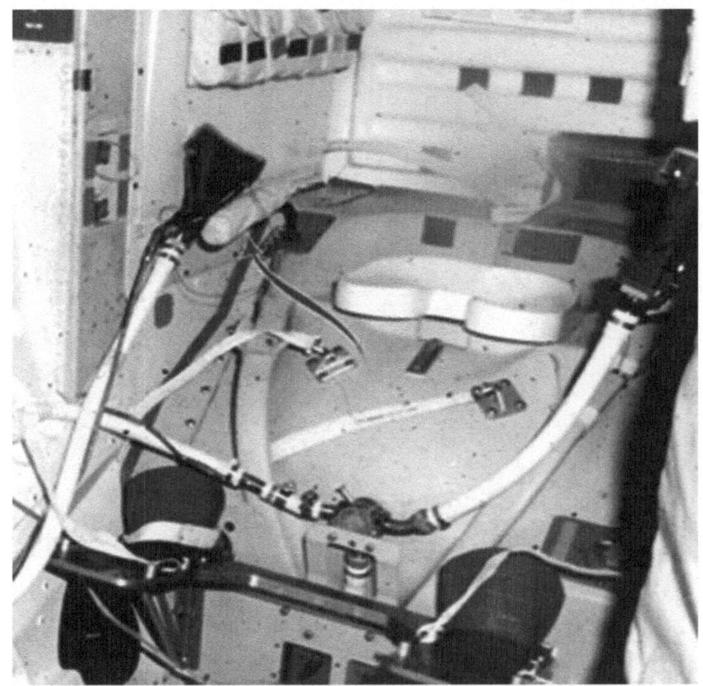

FIGURE 5.9 Shuttle waste management system. (Photo courtesy of NASA)

airflow/suction basis. Figure 5.9 is a picture of the shuttle's waste management system as it is called in NASA lingo or toilet in everyday language. The long tube is for voiding urine. The tube has a personal, color coded, male or female attachment placed on the end before using. Seven of them are in a container in the upper left of the picture. Air flow through the system is activated, creating a vacuum so that urine can be voided through the tube to be passed into a collection and storage unit. Air flow is also used to collect feces. One notable difference in using a toilet in space is that one must be held on the "throne." It would be more than inconvenient to float away while using the facility. The two spring loaded pads, one on each side of the seat, are lifted up, turned inwards, and placed over the thighs. This ensures proper positioning, holding the user in place.

Personal hygiene is primarily accomplished by using damp towels and non-rinsing shampoo. Sponge baths and wipe downs are

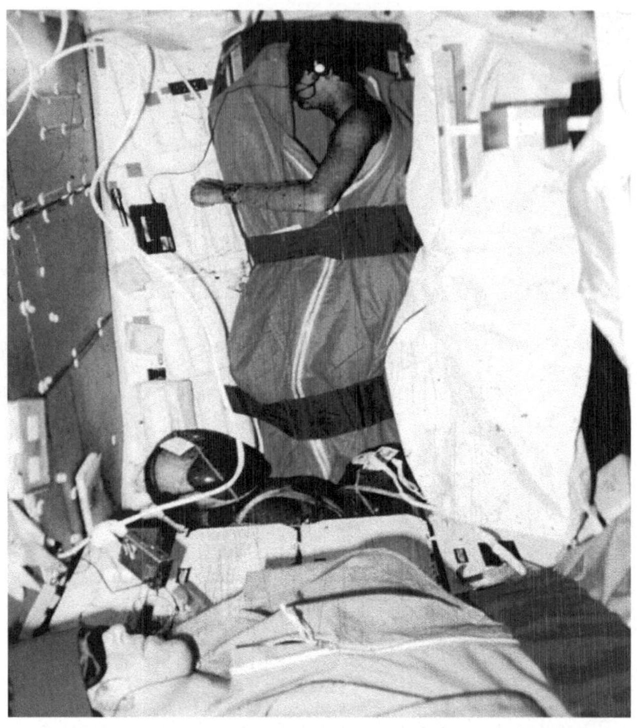

FIGURE 5.10 On the space shuttle, sleeping can be accomplished by placing one's sleeping bag somewhere in the mid-deck and using Velcro straps to secure one's body and head. (Photo courtesy of NASA)

used to complete one's hygiene regime while on the space shuttle and ISS. When it is time to catch up on your rest, sleeping bags can be attached, using Velcro, to different places on the walls of the vehicle. Velcro straps are used across your body so that you do not float out of the bag while asleep (Figure 5.10). All in all, living facilities on either vehicle are functional, but hardly luxurious.

There is another problem to deal with when in space. Using the ISS as an example, it will be in orbit for years with astronauts spending months in space. Living in space is no different than living on Earth. Crew members all generate trash, it seems to be part of our throw away, discard culture. Much waste is left over from food packaging. How do get rid of it? At intervals, as needed, the ISS is re-supplied using the space shuttle or a Russian unmanned Progress space vehicle. Progress does triple duty, serving as a delivery truck, a garbage truck, and an incinerator. After the Progress space vehicle docks with the space station, everything for the crew is unloaded and the vehicle is filled with refuse, including waste water, that has been generated since the last re-supply. Then Progress is released and placed on a trajectory that enters the Earth's atmosphere rapidly. The vehicle and all of its contents are incinerated over a vacant part of the ocean. If there was any remaining debris, it would fall into the water.

This approach works well while in LEO where re-supply and disposal is close at hand. It will not be the answer when we are creating outposts and colonies on the moon and Mars. The policy then will be to reuse and recycle. To be successful, colonies must become self reliant and have a minimal dependency on receiving provisions and supplies from Earth. Space travelers need to remember that great truth, "Throw it away? There is no away!" Our unused trash and refuse will stay with us forever. Waste disposal cannot be ignored just because one is in space. Refuse cannot be ejected to orbit the planet. The space vehicle you are inhabiting might violently crash into bits of refuse that were carelessly discarded. A current hazard of LEO is the many bits of orbiting space debris that could cause tragic consequences if they were to strike a crewed spacecraft. NASA estimates that there are over 9,000 pieces of drifting junk, weighing more than 6,000 tons currently orbiting the Earth. This material originated from earlier space missions, much of it left over rocket boosters and dead spacecrafts. Most of the space junk is currently in a higher orbit than crewed space ships usually go, but eventually those orbits will deteriorate and this refuse will eventually pass through the region where crewed vehicles are. The problem is that the quantity of space debris present in LEO will grow as space travel becomes more common. In your ship, traveling at 17,500 mph in a straight line, it is hard to dodge floating hazards. The common statement in our National Parks is, "Take only pictures, leave only footprints." This could be restated for space voyagers as, "Take only pictures, leave only gases."

6. Tomorrow's Tourist Adventures

We want to build colonies on the moon, Mars, the moons of other planets, and even the nearby asteroids. We want to make space tourism and commerce routine.
Daniel Goldin (1940–Present), NASA Administrator 1992–2001

When space tourism becomes open to the general public, not just the wealthy, the initial trips will be suborbital, up and back with a brief period of ug and an exhilarating view of the curvature of the Earth. The big change in the new world of space tourism is that it will be developed by private commercial companies, not governments. Burt Rutan, who built and flew the first around the world non-stop airplane, and his Scaled Composites Company showed that it can be done. In 2004, they flew SpaceshipOne to space and repeated the feat in less than 2 weeks to win the X Prize (Figure 6.1).

Sir Richard Branson, developer of the Virgin companies including Virgin Atlantic Airways, has come on board and joined with Burt Rutan. They have formed a new space travel company called Virgin Galactic to lead the quest for commercial space flight. A number of these tourist carrying SpaceshipTwo vehicles are under construction. The first model to be completed is called Virgin Space Ship Enterprise or V.S.S. Enterprise. They are designed for two pilots and six passengers. The spaceship will be carried 10 miles up and launched for a brief, minutes long, suborbital space flight. They hope that VSS Enterprise has made several trips into the upper atmosphere. More test flights are planned, and it is hoped that the first paying customers will be taken on suborbital flights in 2012. A picture of Spaceship Two being transported by the White Knight airplane is shown in Figure 6.2.

Although suborbital flights will the first space tourist adventure, Virgin Galactic and other companies that are planning to be active

R.W. Phillips, *Grappling with Gravity: How Will Life Adapt to Living in Space?*, Astronomers' Universe, DOI 10.1007/978-1-4419-6899-9_6,

FIGURE 6.1 SpaceshipOne, the first commercial space vehicle to reach space twice in 2 weeks.

FIGURE 6.2 A photograph of *SpaceshipTwo* being carried to launch altitude by the White Knight or Eve airplane.

in the space tourism industry are eager to reach the next goal. After experience with suborbital flights, the big space tourist agenda will be Low Earth Orbit (LEO). In addition to free fall, floating unhindered by gravitational constraints, viewing an ever changing panorama of Earth day and night will be a major attraction. It is not just fun, it is mind expanding. Most astronauts lose their narrow perspective of looking at or photographing their own particular part of the planet. Instead, they become globalists. The refrain you hear is not "my" city or "my" state but "our" planet. Flying in space does a wonderful job of abolishing parochial perspectives. Perhaps in the future, heads of governments and policy makers should be encouraged, if not forced, to acquire this new vision of Earth as a single entity that develops as a result of going into space. It might be a great help in bolstering peaceful solutions between cultures and countries.

The areas of the Earth that can be viewed will depend on the launch site and the angle of inclination at launch. That establishes how far north and south of the equator the space vehicle flies. If the launch site is located at 30° north of the equator then the path of the vehicle will be at least 30° north and 30° south of the equator. It could also be higher, say 50–60° north and south of the equator. The greater inclination would include views of much of Europe that can't be seen on a space flight with a lower latitude. Whatever the inclination, there are a multitude of great views to record and bring home. Hundreds of thousands of pictures of the Earth have been taken by astronauts and also by satellites with sophisticated cameras. There is only room here to show a few examples of the kinds of scenes that may be recorded during a tourist space flight. Actual pictures will depend on windows of opportunity on that particular trip.

First some pictures from LEO with a circular orbit several hundred miles above Earth. The first was taken from the ISS as it passed over the eastern seaboard and directly over Washington DC, the capital of the United States (Figure 6.3).

Moving west across the United States, the next picture, taken from the space shuttle, is a slanted or oblique view that shows the curve of Earth at the top. The picture includes parts of Colorado, Wyoming and Utah, with a hint of Montana and Idaho near the rim at the top. The brown central part of the picture is Wyoming (Figs. 6.4–6.13).

As space access in LEO becomes more commonplace, it will be possible to go on a more adventurous elliptical orbit. An elliptical orbit gives a broader view of Earth to the observer. One particularly spectacular sight is the line of darkness and light as

FIGURE 6.3 A vertical view of Washington DC from the ISS. The Potomac river flows south east. It is joined by the Anacostia river at the *bottom*. The *white* structure above *center right* is the Capitol (*1*). The White House (*2*) is diagonally NW up Pennsylvania Avenue (*3*) in the center of a dark area. The Washington Monument (*4*) is at the west end of The Capitol Mall. The Lincoln Memorial (*5*) is due west of the Capitol on the banks of the Potomac. The Kennedy Center for the Performing Arts (*6*) is the large white structure just upstream past the Theodore Roosevelt Bridge. The Jefferson Memorial (*7*) is on the tidal basin south of the Washington Monument. The Pentagon is across the Potomac River (*8*). Reagan Airport is at the *lower edge* of the picture (*9*). (Photo courtesy of NASA)

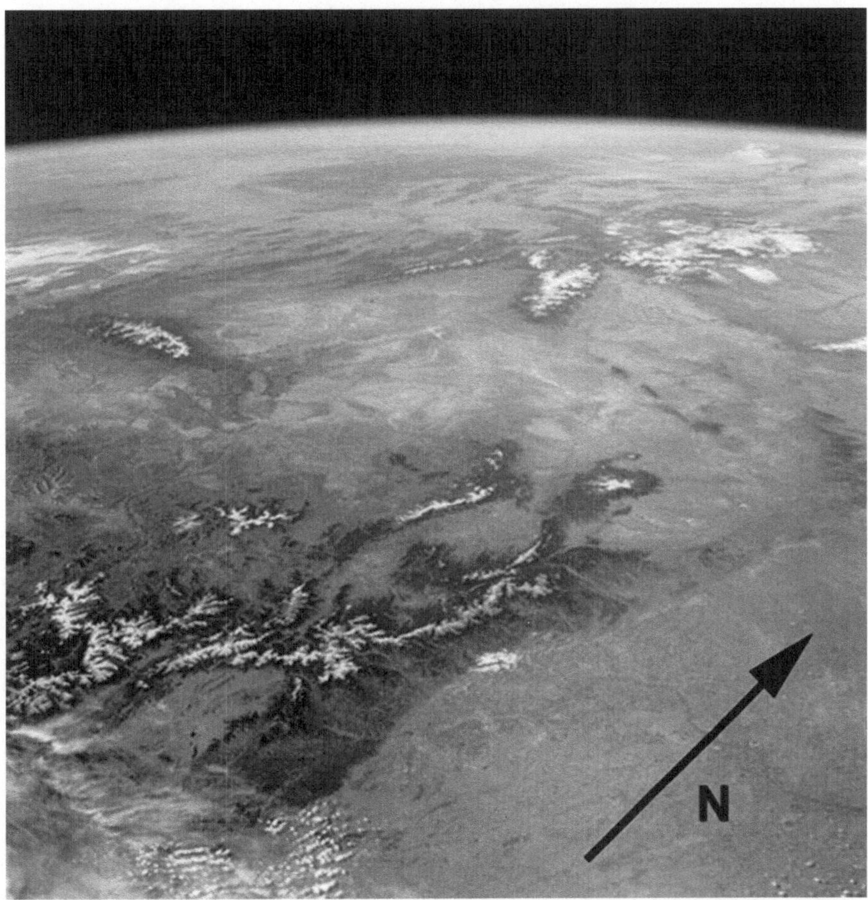

FIGURE 6.4 An oblique or slanted view from the Space Shuttle Atlantis taken October 2002. For a more accurate representation, turn the picture to the left about 45° so that the north *arrow* is straight up. In the foreground is the front range of the Rocky Mountains running north to south. Denver is a *gray* area near the base of the mountains. The South Platte River angles northeast from the mountains across the western portion of the Great Plains. The *round circle of brown* just above the lowest portion of the front range is North Park, Colorado. In the *upper left* is the Great Salt Lake in Utah. To the *lower right* of the lake are the snow-capped Uinta Mountains. The large *brown* area east of the Uinta Mountains in the *middle* of the picture is central Wyoming. The vertical chain of mountains to the north is the Tetons. The snow-capped area in the *upper right* is Yellowstone National Park. (Photo courtesy of NASA)

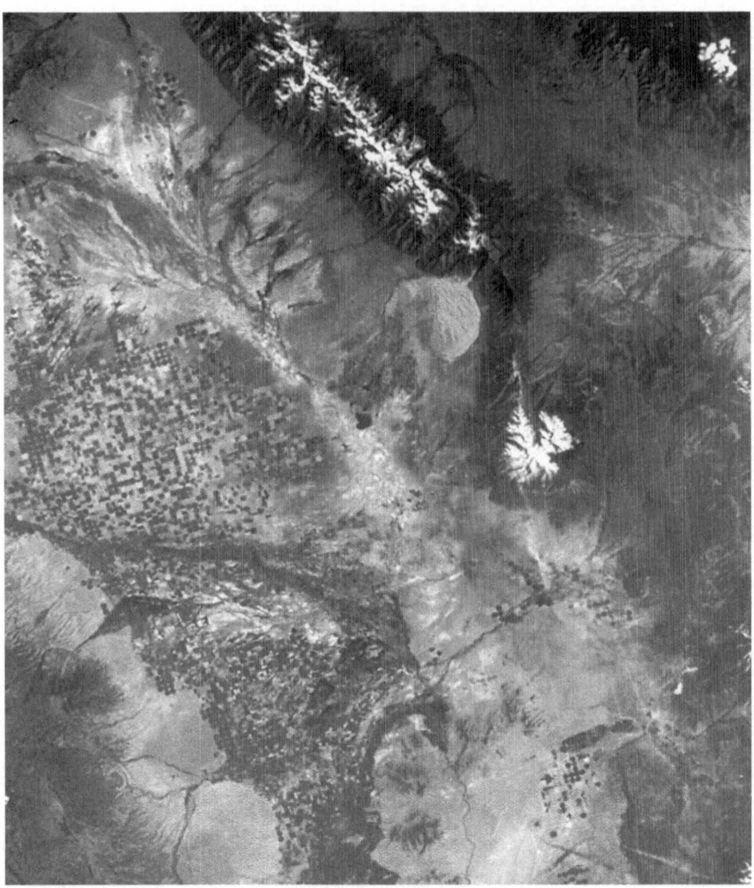

FIGURE 6.5 Great Sand Dunes National Park is the *light yellow circle* south of the snow covered peaks of the Sangre de Cristo Mountains. The individual peak south and a little east of the sand dunes is Little Bear Mountain, one of the tougher over 14,000 ft peaks to climb in Colorado. A number of irrigation crop circles can be seen straight west of Little Bear. In the *lower right* and *upper left* corners, there are two diagonal *white lines*. They are contrails from jet airplanes, both heading NW. Just to the *left* and forward of each contrail is its *dark shadow* on the ground, indicating the position of the sun and time of day (late morning). Photo taken by crew of Columbia June 1991. (Photo courtesy of NASA)

FIGURE 6.6 San Francisco Bay area as seen from a space shuttle. The Golden Gate Bridge connecting the city with Marin County is to the *lower left* in the picture. The Oakland Bay Bridge is below and to the right of the Golden Gate Bridge. Alcatraz Island is the small dot north of San Francisco. San Pablo Bay is the circular body of water at the *top*. The Sacramento River terminates as it enters San Pablo Bay. The Diablo Mountains are on the *lower right*, east of San Francisco Bay. (Photo courtesy of NASA)

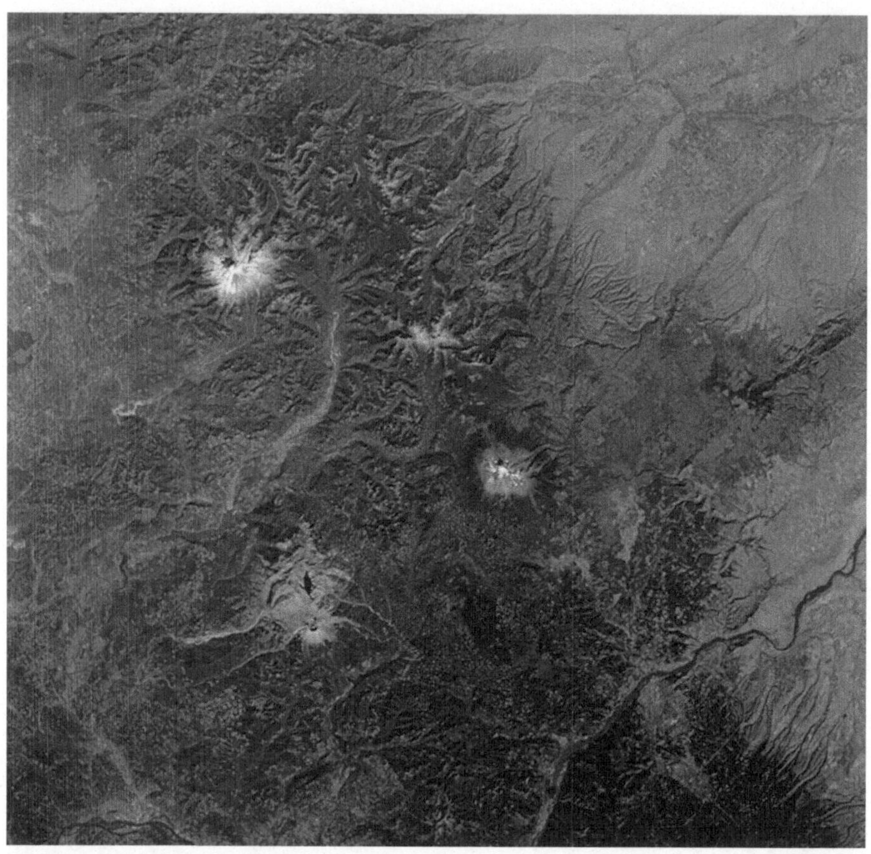

FIGURE 6.7 View from Space Shuttle Atlantis, October 2002. North is towards the *upper left* corner. The picture includes four volcanoes in the Cascade Mountains. The green forests of the Cascades are one of the main features of the picture. Mt. Rainier is at the *top*. In the plains on the *left* edge is Seattle. On the east are the semiarid plains of the Columbia River basin. Mt. Adams is to the *right* of *center* and Mt. Hood in the *lower right*. The Columbia River gorge is north of Mt. Hood. The river flows mostly west. Mt. St. Helen is SW of Mt. Adams. The remains of its cone can be seen at the *bottom* of the *brown* area. When Mt. St. Helens erupted in 1980, it blew out the north side of its cone and destroyed many square miles of forest. The remnants of the mountain were too low to retain snow in October. (Photo courtesy of NASA)

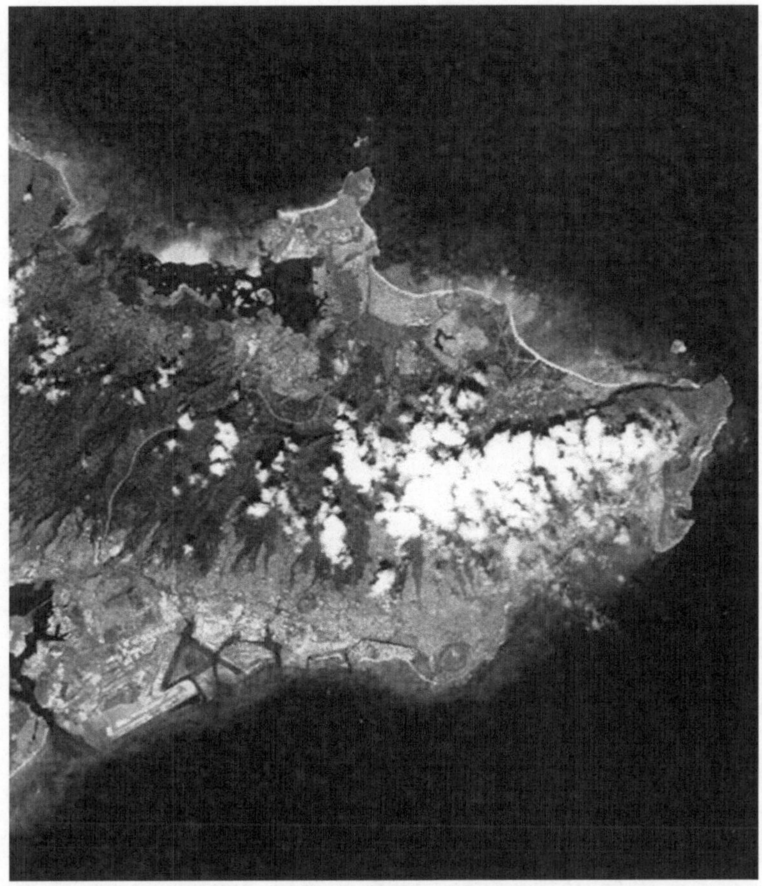

FIGURE 6.8 Honolulu, Hawaii, on the island of Oahu from the Space Shuttle Endeavor, February 2000. Diamond Head, an extinct volcano, is just to the *right* of *center* as an ellipse. The irregular *white line* to the left of the volcano is Waikiki Beach. Left of the beach is the Honolulu Airport. Pearl Harbor is at the *far left*. (Photo courtesy of NASA)

FIGURE 6.9 Picture taken by the crew of the Space Shuttle Atlantis, September 2000.The Sea of Galilee is in the *lower center* of the picture. At the *left* is the Mediterranean Sea. The port of Haifa is west of the lake in a bay on the coast. The Golan Heights overlook the lake on the east. (Photo courtesy of NASA)

FIGURE 6.10 Photo by the crew of the ISS of Mount Etna erupting on the Island of Sicily in the Mediterranean Sea. Mt. Etna has the longest period of eruptions that have been recorded. The *white* smoke, part way down the left slope, is due to forest fires started by hot lava flowing down from the eruption. (Photo courtesy of NASA)

FIGURE 6.11 Hurricane Lili in the Gulf of Mexico October 2002, as seen from the ISS. At the time of the picture, the eye was over 15 miles wide. (Photo courtesy of NASA)

FIGURE 6.12 Cape Cod, Plymouth, MA, and Boston, MA, in February 2000, as seen from the Space Shuttle Endeavor. The near horizontal line at the *bottom* cutting across the narrow part of the Cape is the Cape Cod Ship Canal that turns the Cape into an island. Provincetown and sand dunes are at the end of the Cape. Plymouth, the site of Plymouth Rock, and its breakwaters are at the top of the bay just below *center*. Boston is near the *top* of the picture. (Photo courtesy of NASA)

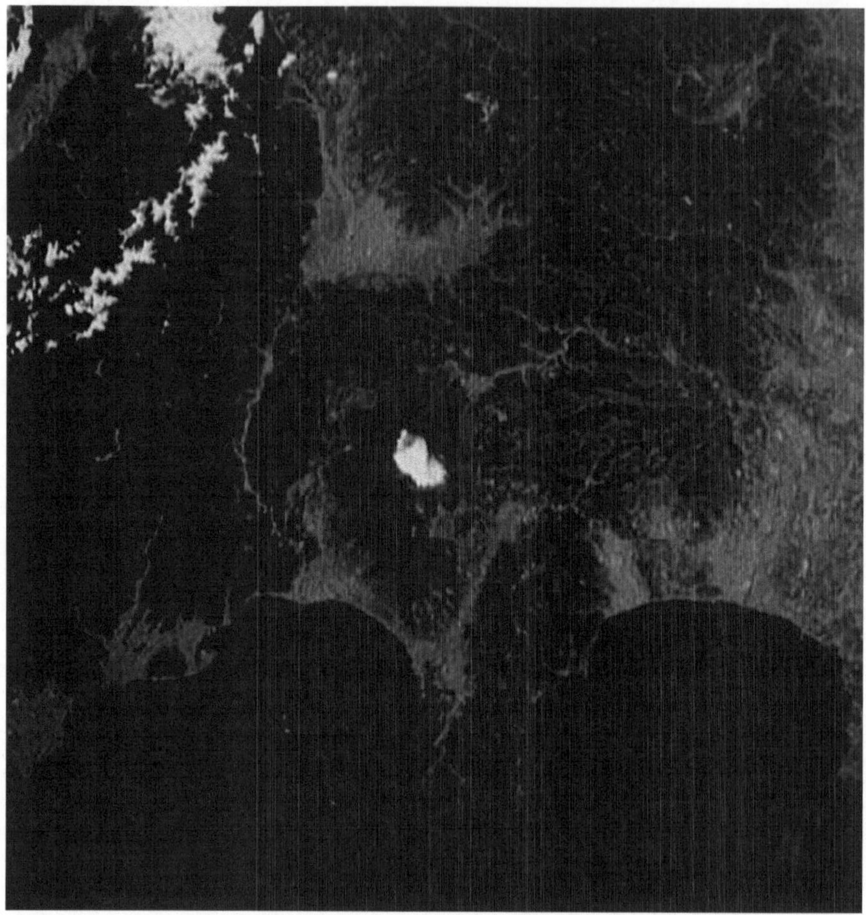

FIGURE 6.13 Mt. Fuji is the snow capped peak in the *center* of the picture. It is located on Honshu Island, the largest Island of Japan. The city of Tokyo is located on the flat plain in the *lower right*. (Photo courtesy of NASA)

dawn breaks or nightfall occurs on a continent (Figure 6.14). At perigee, the low point of the orbit, the spacecraft will be up only a few hundred miles. At the upper end of the picture it is a bit over 2,000 miles above the surface, which is the apogee. This altitude would provide a vertical view of most land masses on the Earth. An example of such a view is shown in Figure 6.12 that includes much of North America (Figs. 6.15–6.19).

FIGURE 6.14 An elliptical orbit with the apogee, or furthest point from the Earth, about 2,000 miles above the surface, as compared to the perigee, or low point, at 200 miles up at the *bottom* of the picture.

The next big step for space tourism after suborbital flights, LEO and elliptical orbit flights will be to journey to the moon. Before that can happen, we must first establish an outpost which is planned to be developed in the 2020s. It will be designed to support exploration and scientific research. The outpost will grow, diversify, and become a colony. One way for the colony to help support itself will be to develop tourist facilities. Full scale tourist trips won't be started until colonies are constructed that have a significant infrastructure, as well as facilities that will accommodate visitors/tourists. A lunar hotel would include not only living

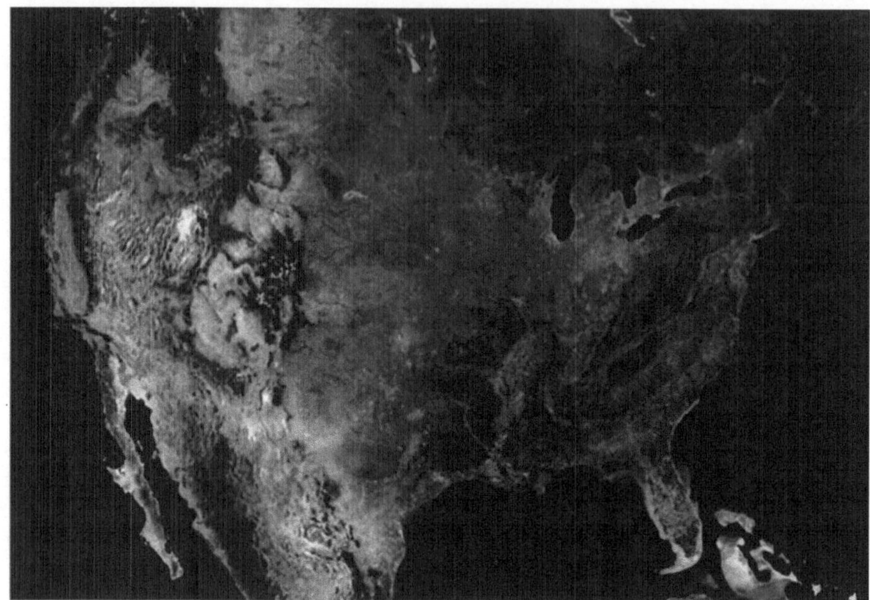

FIGURE 6.15 A view of North America including the United States and parts of Canada and Mexico. The picture is a composite without cloud cover. A view such as this provides little evidence of civilization, development, or national boundaries. One exception, the *lighter color* that is the Chicago urban area can be seen at the southwest edge of Lake Michigan. The differences in climate, particularly annual moisture, between the eastern and western parts of the continent are very obvious. (Photo courtesy of NASA)

quarters, but inside recreational facilities as well as an opportunity to walk on the moon. The simplest outing would be similar to the early Apollo visits. You would need to wear a lunar space suit and a life support backpack to provide an oxygen supply and carbon dioxide removal system. Walking or hiking trips may last several hours in the vicinity of the colony. Even a lunar golf course is a possible activity. Alan Shepard started it by hitting several golf balls a considerable distance while on the moon during the Apollo 14 mission. A golf course on the moon won't have a green, but there can be lots of hazards. Rocks and boulders are plentiful, but

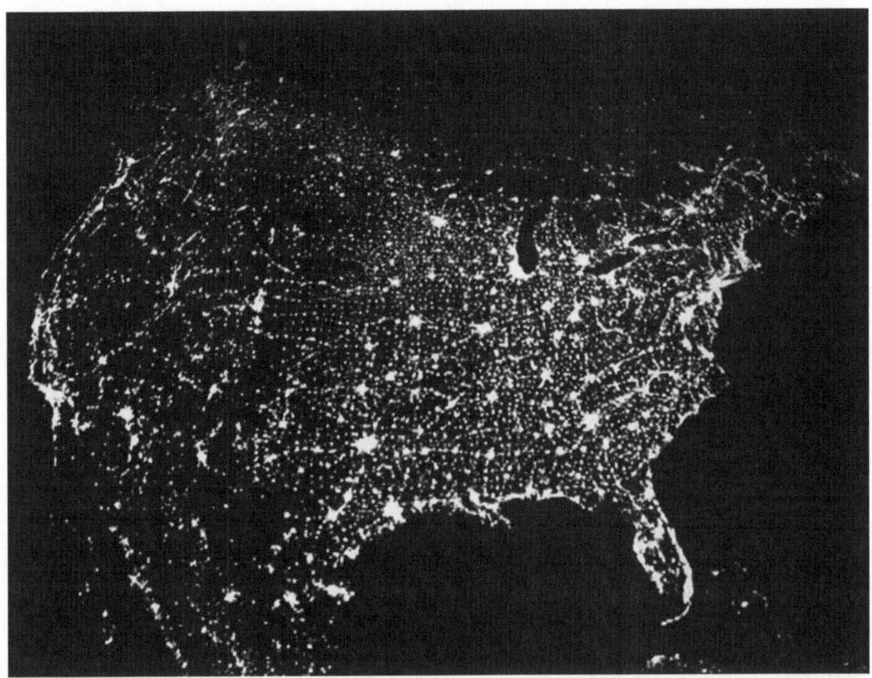

FIGURE 6.16 A view of North America at night that is similar in coverage to the previous daytime picture. All of the major metropolitan areas of the United States can be seen. The New York/New Jersey metropolitan region as well as Boston and the Florida eastern and western seaboards are prominent on the east coast. Chicago/Milwaukee stand out as a single entity as do Houston/Galveston and Dallas/Fort Worth. It is easy to see from this picture that the populated areas of the country in general, not just the major cities, are light polluted. There are no indications of country borders, although cities on the Canadian side of the Great Lakes are obvious as are some cities in Mexico. The upper portion of the Sea of Cortez indicates the approximate border between the United States and Mexico. There are not many places left where one can find truly dark skies in the eastern and central parts of the continent. (Photo courtesy of NASA)

no roughs with trees and tall grass. You may come back to Earth having hit a golf ball further than Tiger Woods ever did.

Advanced models of the Apollo lunar rovers will be capable of carrying several passengers and perhaps a guide to lead the adventure out from the moon base hotel. Space suits will still be necessary. The freedom to change direction and explore newer areas

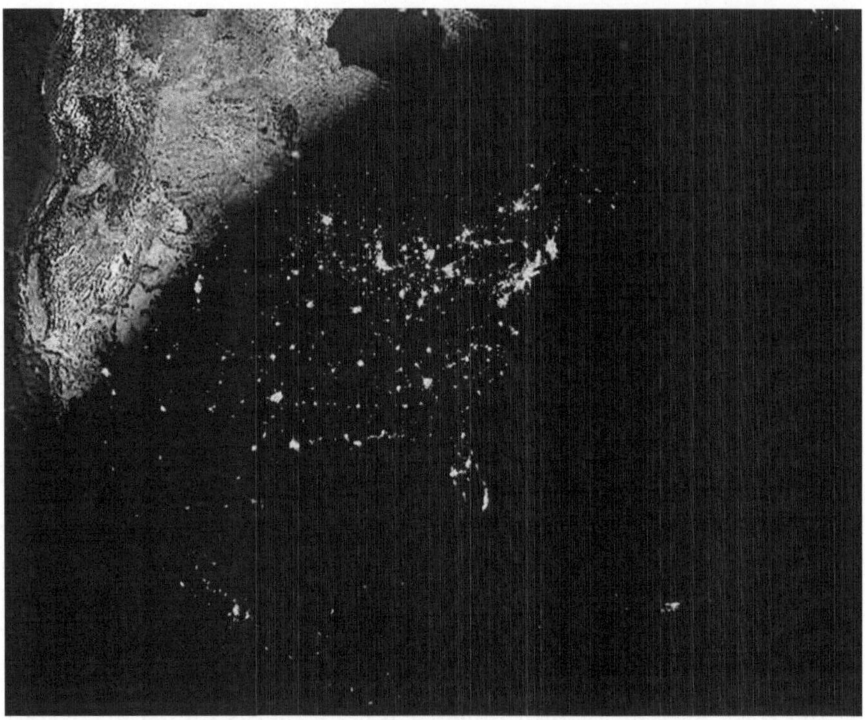

Figure 6.17 This view of North America is a combination of the two previous pictures. The line of darkness, called the terminator line, curves northeast from near San Diego in the south to just west of Denver and west of Minneapolis/St. Paul. The terminator line continues across the southern portion of Hudson Bay in northern Canada. (Photo courtesy of NASA)

will add to the trip. A more advanced tourist outing will be in an enclosed and pressurized lunar vehicle in a suit free environment. It will be possible to travel for a number of hours and to explore unique lunar sites.

A lunar hotel and tourist facilities may seem like only a dream for the future, but it was too not long ago that traveling to space was only a dream. It is now a reality for Russia, the United States, and China. Other countries, notably Japan, India, and a consortium of European countries plan to develop space programs that include human exploration of the moon. A number of private commercial entities are working to join this group to support science, exploration, and tourism.

But why should we go to the moon at all, much less as tourists? Perhaps an answer can be found in the words of the Apollo

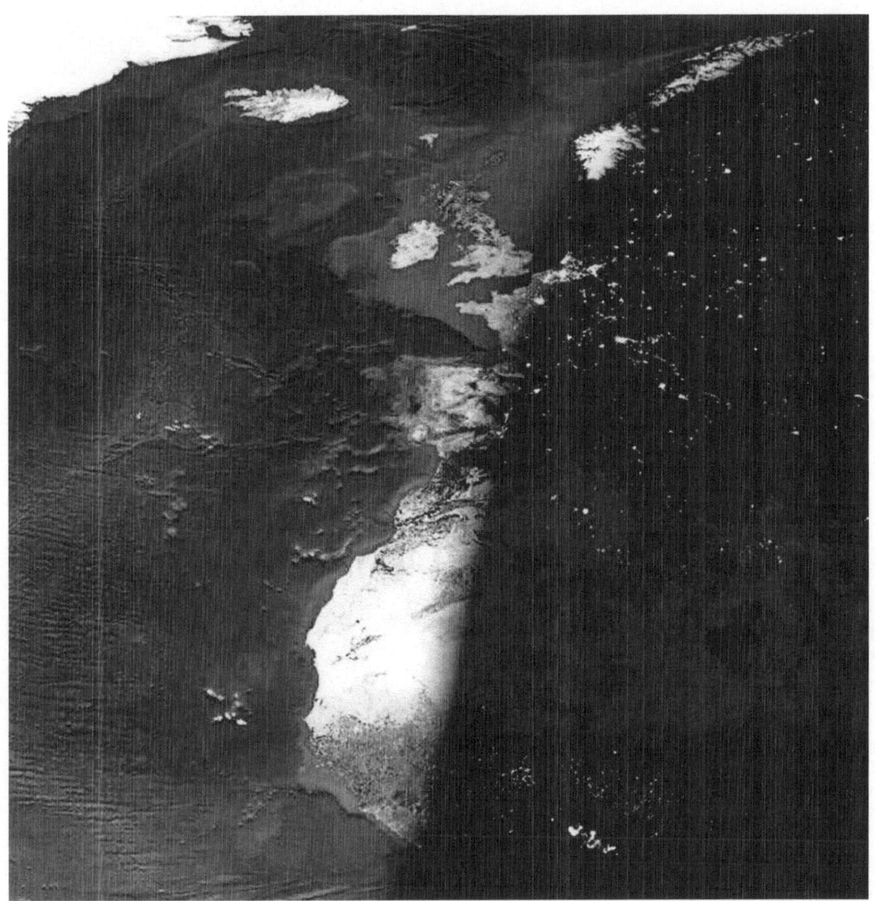

FIGURE 6.18 Western Europe and Africa in an oblique view that shows Iceland west of Scandinavia and a small corner of Greenland in the *upper left* portion of the picture. The near junction of Europe and Africa across the Straits of Gibraltar is almost exactly in the *middle* of the picture with the darkness of the Mediterranean Sea to the east. Although there is snow in Norway, Sweden, Iceland and Greenland the countries of England, Scotland and Ireland are still green. The scarcity of lights in western Africa outline the barren Sahara Desert in the dark side. The Canary Islands are west of the southern portion of Morocco where the Sahara Desert reaches the Atlantic Ocean. (Photo courtesy of NASA)

pioneers who have been there. They are vocal advocates of exploring and opening the moon to Earth dwellers:

> As I stand out here in the wonders of the unknown of Hadley, I sort of realize there's a fundamental truth to our nature. Man must explore, and this is exploration at its greatest.
>
> Alan Shepard, Apollo 14

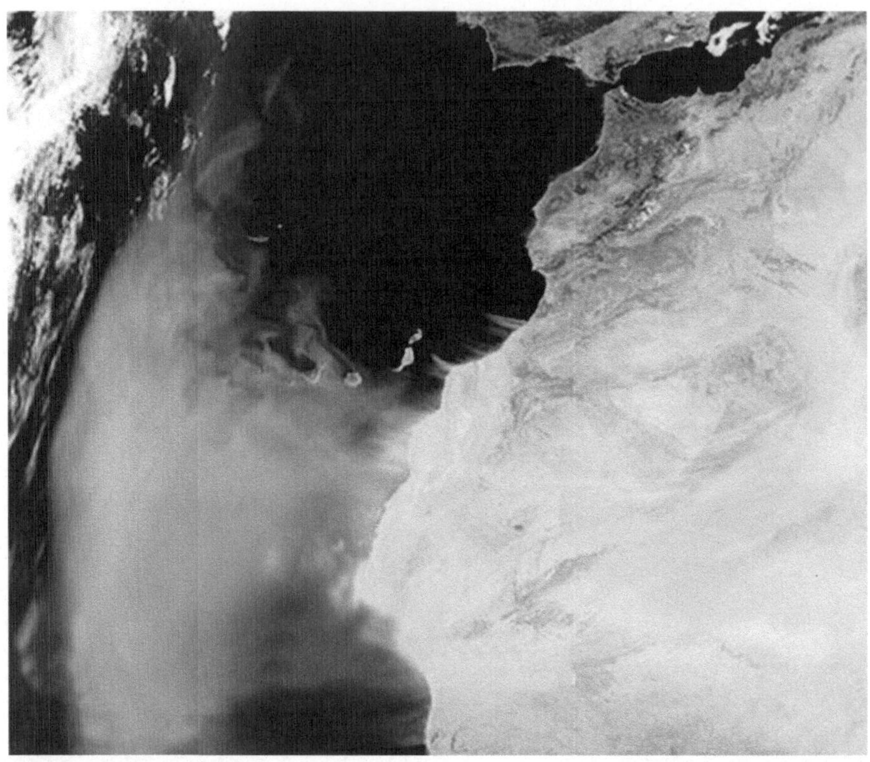

FIGURE 6.19 A sand storm blowing out of the Sahara Desert in northwest Africa across the southern portion of the Canary Islands and on out into the Atlantic Ocean. The southern tip of Spain, Straits of Gibraltar, Morocco, and the western Mediterranean are clearly visible. (Photo courtesy of NASA)

> It's like trying to describe what you feel when you are standing on the rim of the Grand Canyon.
>
> Alan Bean, Apollo 12

> That America's challenge of today has forged man's destiny of tomorrow.
>
> Gene Cernan, Apollo 17

> For when I look at the moon I do not see a hostile, empty world. I see the radiant body where man has taken his first steps into a frontier that will never end.
>
> David Scott, Apollo 15

Isn't that something? Magnificent sight out there.

Neil Armstrong, Apollo 11

Magnificent desolation.

Buzz Aldrin, Apollo 11

We went to the moon as technicians, and returned as humanitarians.

Edgar Mitchell, Apollo 14

It's human nature to stretch, to go, to see, to understand. Exploration is not a choice, really, it's an imperative.

Michael Collins, Apollo 11

In my own view, the important achievement of Apollo was a demonstration that humanity is not forever chained to this planet, and our visions go further than that, and our opportunities are unlimited.

Neil Armstrong, Apollo 11

Suddenly, from behind the rim of the moon, in long, slow movements of immense majesty, there emerges a sparkling blue and white jewel, a bright delicate sky-blue sphere laced with slowly swirling veils of white, rising gradually like a small pearl in a thick sea of black mystery. It takes more than a moment to fully realize this is Earth....home.

Edgar Mitchell, Apollo 14

The distance to travel to reach the moon is 238,000 miles as compared to LEO several hundred miles above Earth's surface, but it does not have to be a long trip. On Earth to moon missions during the Apollo flights, the astronauts went first into LEO at 17,500 miles per hour. After several orbits spent checking the spacecraft's systems, they fired the escape rockets to increase their speed to 25,500 mph and journey to the moon, a voyage of less than 10 h.

In the future the trip may be made in a single step. You will leave the Earth's surface and accelerate to escape velocity. The journey to the moon can be completed in less than a day. It will be a great trip watching the moon grow larger and larger while in the opposite direction Earth will be rotating so that you will have the opportunity to enjoy distant views of our planet. Once in the moon's vicinity the vehicle will slow and go into a lunar orbit prior to landing. As the lunar express vehicle goes to the back side of the moon, the lunar voyagers will see a picture like that shown in Figure 6.20. After landing on the surface near the tourist living facilities, that are not yet designed, the voyagers will move into their new accommodations. This will be a unique opportunity to enjoy the slight gravity field and to spend time observing and exploring a brand new domain unlike anything available on Earth (Figs. 6.20–6.23).

A fantastic future is in development. We will soon be able to experience new vistas that were impossible to even consider just a few years ago. It boggles the mind to imagine what new opportunities will be available to our grandchildren and their grandchildren as they mature.

FIGURE 6.20 Back side of the moon that is not visible from Earth taken by Apollo astronauts. (Photo courtesy of NASA)

FIGURE 6.21 A picture taken by Apollo 15 astronauts of a fresh meteor crater in the foreground and part of the Hadley Delta Mountain in the distance. (Photo courtesy of NASA)

FIGURE 6.22 Lunar landscape with Harrison Schmidt of Apollo 17 and a large boulder in the foreground. Picture taken by Gene Cernan near the Taurus Littrow landing site. A large valley with lunar hills is in the background. (Photo courtesy of NASA)

FIGURE 6.23 Earth rise as seen from lunar orbit during the Apollo 11 mission, July 1969, the first mission to land humans on the surface of the moon. The picture was taken by Michael Collins in the Command Module. (Photo courtesy of NASA)

Part II
Life in Space

7. Taking Your Body to Space

WATER IN SPACE
Water in and water out
That's what life is all about.
When to space we finally go,
It provides a greater flow.
Our lower legs no longer swell,
This new design, it suits us well.
Trouble comes when we're first back
Blood for the brain, there's a definite lack.

Anonymous

Part A: Hello Bird Legs

We are water filled sacs. In an adult human, water represents about 60% of our weight. At birth, infants are about 77% water, but slowly dry out over the years as they age. Our water content may be as low as 45% in old age. Dry matter is the smaller portion of your heft for most of life.

Here on Earth, water follows gravity regardless of our desires. In our bodies, it runs downhill and collects in legs and feet. A common occurrence when we sit for long periods in an airplane, a car, or a movie is that our feet and lower legs are swollen. They fill with water that has drifted down. We are built that way. If you take your shoes off for a while, it may be hard to put them back on. All of that sinking water enlarges the feet and lower legs. Sometimes we become varicose.

Unfortunately in much of the high-tech societies, boob tubes and all, we have become a sedentary lot. In the past, we walked and exercised to go from one place to another. That simple activity used to keep water and blood from collecting in our lower limbs. Try it, it still works.

R.W. Phillips, *Grappling with Gravity: How Will Life Adapt to Living in Space?*, Astronomers' Universe, DOI 10.1007/978-1-4419-6899-9_7,

The majority of body water is inside cells, but there is almost as much between cells and in the blood. Water in the body moves easily and rapidly from one spot to another. The fluid in blood vessels and the water between cells are called extra-cellular fluid (ECF). The ECF rapidly changes when you enter space. As soon as you get into LEO, water is no longer held down by the force of gravity. It moves rapidly throughout your body and becomes uniformly distributed. You also don't look the same because you lose about a quart of water from each leg. Your legs get skinny, while your face becomes swollen and flushed as blood and water collect in your upper parts (Figure 7.1).

When you first get to space your nose and sinuses become congested, your eyes water, and bird-like legs develop. Your water balance is out of kilter, and body control systems take over. You begin to adapt to an abundance of water in your upper body. So what do you do? Slow down the intake and keep up the output.

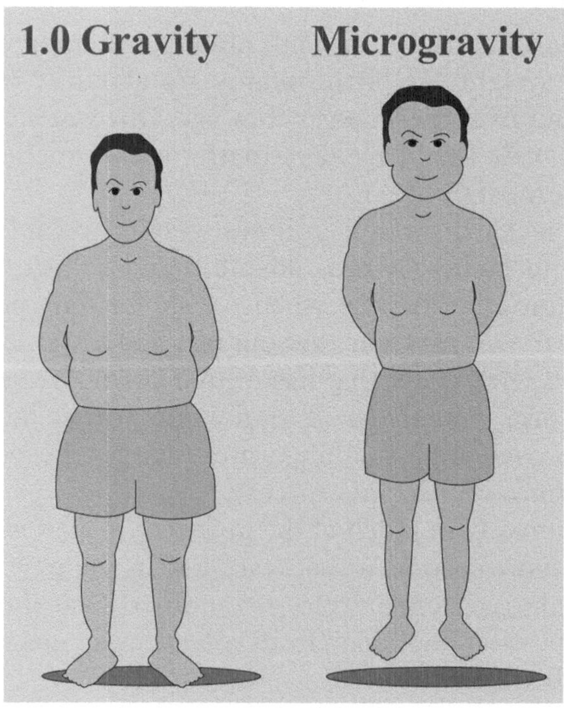

FIGURE 7.1 Water no longer accumulates in your legs when you reach space, but distributes more equally throughout your body.

That's just what happens, you drink less and continue to form urine. In a few days, you arrive at a new, space adapted, body water balance. One with less total water than you had on Earth, but a normal supply of water in the upper body. The change is mostly in water outside the body's cells. There can be a big decrease in total blood volume, up to 25%. That decrease in blood volume creates another problem. A tightly regulated system in the body is the percentage of red blood cells, rbc or erythrocytes, in the blood. When the water content of your blood goes down during space flight, the total number of rbc remains the same. There are more cells per unit of volume. Too many! Blood flow in small vessels is sluggish and some capillaries could become plugged. Your body needs to adapt.

The body control system that regulates rbc production starts in the kidneys. In addition to removing wastes from the body by urine formation, the kidneys recognize and control the concentration of rbc circulating in the blood. When the kidneys sense rbc increase or decrease, they alter the secretion of a hormone called erythropoietin (EPO) that causes the bone marrow to manufacture and release more rbc. In space, when your blood volume goes down due to water loss, the rbc concentration increases and your kidneys start the adaptation by decreasing EPO formation and secretion. It is a finely tuned and responsive system that keeps the percentage of rbc in your bloodstream under tight control (Figure 7.2).

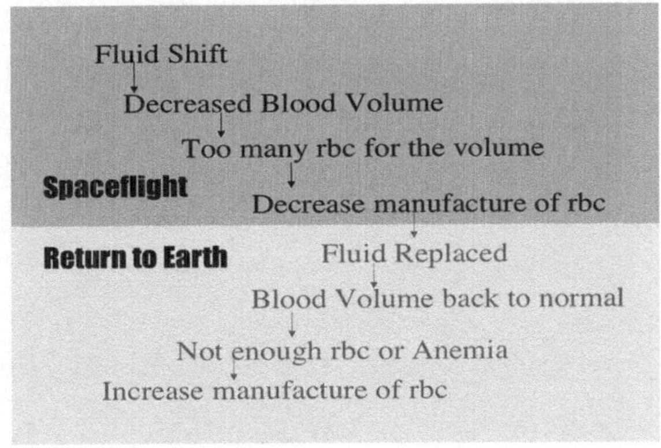

Figure 7.2 The fluid shift initiates a cascade of adaptive changes that result in a decrease in blood volume followed by a decrease in the total number of red blood cells during space flight.

As a result, while in space, at least on a short term mission of weeks, your body will decrease production of new rbc. Reducing the number of rbc in the blood is a slow process because the average lifespan of a normal rbc is about 100 days, so a relatively small number of total cells are lost each day.

Figure 7.3 shows the changes in erythropoietin concentration in the blood as a result of traveling to space for a few days and then returning to Earth. EPO secretion by the kidneys is reduced shortly after entering space and then increases immediately upon restoration of a normal blood volume after landing, when your body rehydrates. It is a very predictable adaptation to space and then re-adaptation to Earth's gravity. When you return from a space voyage, the body rehydrates and blood volume increases. As that happens, there are not enough circulating rbc, a condition described erroneously as space anemia in the early days of space flight. Anemia *only* develops post flight after rehydration, not during space flight. After return from space, increased EPO means that new rbc will soon be entering the blood and space travelers will no longer be anemic.

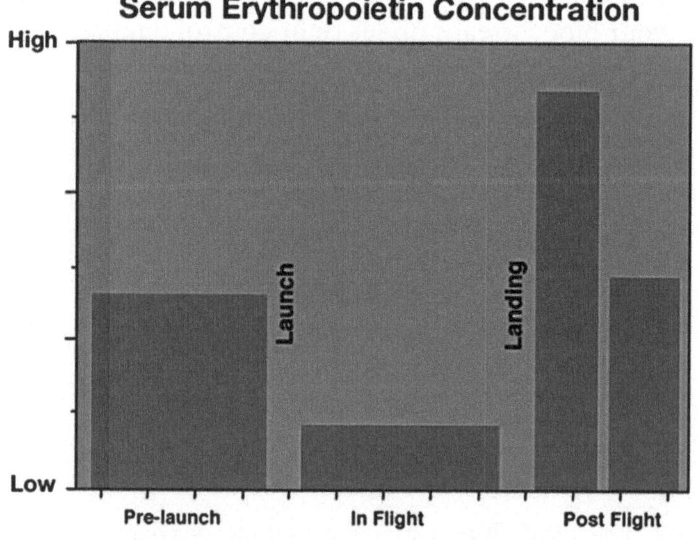

FIGURE 7.3 Erythropoietin (EPO), a hormone from the kidneys, controls the rate of red blood cell (rbc) formation.

The problem with adaptation to space and re-adaptation to Earth gravity would be eliminated if we could only let our bodies know that we are just off on a short trip to this new environment. No reason to make a fuss over a temporary change in location. Unfortunately, Mother Nature doesn't work that way. She sees a problem and immediately starts the remedy. It's good to have a knowledgeable benefactor looking over us.

When you are μg adapted, your total body water and blood volumes are both decreased. You are in great shape to be in space, but no longer as fit to live on Earth. When it is time to return home, your body must readapt to the Earth's gravitational field. Unfortunately, the decrease in blood volume while in space creates another problem when you return. Water movement is just the reverse of the adaptations that happened as a result of entering space. They are illustrated in the bottom section of Figure 7.4. When you are exposed to Earth gravity, water/blood moves back into the lower portion of the body. Overall water was lost during adaptation to space, so there is not enough water left in the body to completely

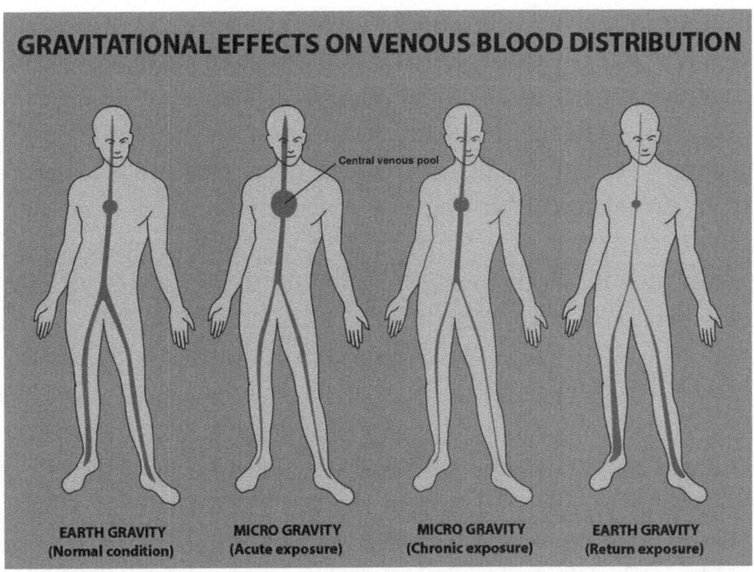

FIGURE 7.4 Under normal Earth gravity conditions, there is a large quantity of venous blood stored in our legs. It shifts upward upon reaching μg. We adapt by reducing the amount of blood as well as other fluids in our bodies. Upon return to Earth gravity, blood flows back down to our legs.

fill all of the body's water pools. This is not a good situation. At the top of the body's water/blood pool in an erect human sits the brain in its bony case. It's the first part of your body to feel the deficit and that can cause trouble. On Earth when you stand up, there is a significant gravity gradient from your head down to your feet. Since water flows down in response to gravity, your head and brain are the first portions of your body to have a water decrease in the early hours after returning from space flight. Fortunately, blood volume is restored fairly rapidly by drinking fluids after landing .

The decrease in blood volume that occurs while in space means there is insufficient blood to fill all of your body pools when you first return to Earth. Earth gravity ensures that the veins in the legs are filled again. As this happens, there is often not enough blood to adequately supply your brain. Astronauts and cosmonauts often suffer from a condition with the fancy name of "orthostatic intolerance." Orthostatic means "caused by standing erect." So returning astronauts are intolerant of standing. They often feel lightheaded and are more susceptible to fainting, particularly if they stand suddenly. They also have difficulty successfully completing a "stand test" that requires that they stand motionless for a brief period of time. This unsteadiness is temporary as they rapidly replace the lost fluid volume and begin to increase rbc formation on their return to an Earth adapted state.

Another countermeasure or method of preventing the μg water change effects is an interesting device called a Lower Body Negative Pressure (LBNP) suit. One of these, a Russian version, is pictured in Figure 7.5. Similar devices have been used in both American and Russian programs although the design of the suit varies a bit. The principle behind this approach is to create a vacuum around the legs. The plan is that the vacuum would pull water from the upper body back into the legs and, in addition, increase blood volume. The water/blood, in turn, will stretch vessels and tissues in the legs and more water will be retained in the lower body after the vacuum in the suit is turned off. Thus, the body would not be as water deficient as before. It's an innovative idea and worth trying, but it hasn't been very effective. I am sure Mother Nature laughed mightily over this one.

The use of LBNP is often combined with drinking more fluids containing salt to increase body water. The results of LBNP and fluid loading may be a short term fluid increase but no substantive effect on overall water balance while in space. In addition, the LBNP suits are quite cumbersome and the astronauts or cosmonauts don't like to wear them.

In addition to the clumsy suits, there has to be a tight band around the waist to prevent the vacuum from sucking on the upper part of the body. Re-hydrating space voyagers as soon as possible after their return is the most effective approach.

When there are microgravity-induced changes in the quantity of body water and in its distribution, there is another problem to deal with. Between each of the vertebrae in the back, there is a small fluid filled disc that acts as a cushion to keep the bones from rubbing each other. Most people are aware that there are these

FIGURE 7.5 Lower Body Negative Pressure (LBNP) suit. One suspects that the only reason that the cosmonaut pictured is smiling is because his LBNP suit is not turned on. (Photo courtesy of NASA)

discs that can rupture or slip and cause severe lower back pain. Rupture or slippage is unlikely to be a problem in space flight as there is nothing heavy to lift or strain against. However there are changes. Normally the discs in your lower back or lumbar region are compressed and flattened as the weight of the whole upper part of the body rests on them. That's a lot of weight for a small water-filled balloon. In space where there is no weight pushing down, these discs in the lower back absorb fluid and become fatter. When this happens, you grow in height, several inches taller for most astronauts (Figure 7.6). The growth in height stretches tiny muscles between individual vertebrae, particularly in the lumbar region. Lower back pain due to muscle stretching will develop and continue for several days. Lower back discomfort is the most common complaint of space voyagers in the first few days of flight. All of these changes reverse when astronauts return to Earth. Even without coming back to Earth's gravity field, the muscles will

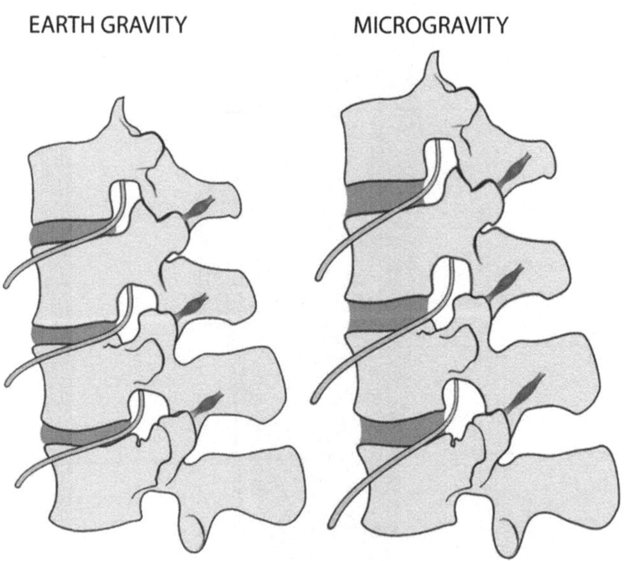

Figure 7.6 During microgravity exposure there is no upper body weight pushing down on the lumbar vertebrae. Fluid accumulates in the invertebrate discs. As the discs expand, small muscles between each vertebrae are stretched, causing discomfort. The result is that you grow several inches in height while traveling in space. (The artistic credit is to Dennis Giddings)

accommodate to their new stretched position and the back pain will fade away.

The rapid changes in water and blood distribution and composition that occur when you travel to space are quickly reversed when you return to Earth. Understanding these changes gives you a clear insight into how flexible your body systems are and how automatically we respond to environmental changes. Similar changes, but to a lesser degree will occur when we return to the moon for longer stays and eventually expand space ventures to our neighboring worlds. Changes in the less fluid parts of our bodies will also occur. They are slower to develop, but are still dramatic.

Part B: Pump and Pipes

THE AMAZING LIVING PUMP
"YOUR HEART"
ABOUT THE SIZE OF YOUR CLENCHED FIST
BEATS/MINUTE = 70
BEATS/LIFETIME (70 years) = 2,575,440,000
BLOOD PUMPED/MINUTE (at rest) = more than 5 quarts
BLOOD PUMPED/DAY = 18,000 gallons
BLOOD PUMPED/LIFETIME = 46,000,000 gallons
This amount of blood would fill a tank 10 feet deep,
100 feet wide and over one mile in length.

Robert W. Phillips (1929–Present), Veterinarian, Physiologist, Author

Just like our homes, our bodies have systems that make everything run smoothly. You need a plumbing system for fluid transport and an electrical system to operate the machinery. The plumbing system is much more highly evolved and better controlled than any house plumbing system ever imagined.

The pump, your heart, increases or decreases flow as conditions change. It does this by pumping more or less blood per beat and by increasing or decreasing the rate of beating. The pipes, blood vessels, also have a controlling function. They are not constant like the pipes in our homes. They can enlarge and carry more blood or constrict and carry less. The pump and pipe system is always changing, depending on the rate of activity in your body. With this flexible control system, the amount of blood going to different sections of your body is continually changing, based on needs of the different parts. If you skip a meal, or even two, blood flow to the gastrointestinal tract will be decreased because it has less work to do. Exercise increases blood flow and pump (heart) output. Most of that increase goes straight to the exercising muscles. The ability to manage and control a variable delivery system is the big difference between household plumbing and the plumbing in your body. We call the whole operation, the cardiovascular system. Cardio stands for heart and vascular for blood vessels. It is a lot more detailed and complex than the few pipes in your house. How does the cardiovascular system work? How does it change when you go to space?

Your heart has four chambers and it serves as the pump for this physical transport system. The two chambers on the right side supply blood to the lungs and the two on the left supply all of the rest of the body. With each beat, your heart sends equal amounts of blood in the two different directions. Equality is essential or there would be a buildup and a backlog of blood on one side or the other. On the right side, venous blood that has a low oxygen and high carbon dioxide (CO_2) content is sent next door to your lungs. This is a low pressure system as it travels a short distance but the flow rate is high. In the lungs, the red blood cells (rbc) release CO_2 that crosses a membrane into tiny terminals at the end of your lung airways. The CO_2 is exhaled from the body as you breathe out. As you breathe in, oxygen from inhaled air crosses that same membrane and enters your blood stream where it is bound by hemoglobin in the rbc that have just given up CO_2. The oxygenated rbc are transported to the left side of your heart and then out to the rest of your body. The left side of the heart is a high pressure system. Blood must be pumped from the heart to all of your body's cells, from nose to toes.

When blood pressure is measured, two numbers are obtained. They represent the pressure inside the arteries leaving the left side of the heart. The first number is called systolic. It is the highest pressure, immediately at the end of a heart beat. The second number, diastolic, is the low pressure that occurs between each beat of your heart.

The gas exchange that occurs in your lungs is reversed in capillaries all over the body. They are the smallest of blood vessels and connect the small arteries coming from your heart with the small veins leading back toward your heart. In the capillaries, oxygen is given off from the rbc, passes out of the blood stream and into adjacent tissue cells to support their metabolic functions. Nutrients derived from food in the stomach and the intestines also leave the capillaries and enter the body's cells. Carbon dioxide leaves these cells to be carried through the venous system to the right side of your heart, on to your lungs, and out into the atmosphere.

The basic function of your heart, lungs, and blood vessels is to transport oxygen and nutrients to cells and carbon dioxide from cells to your lungs. Your kidneys also receive arterial blood from your heart and, in effect, strain it to remove unwanted substances. In addition your liver can remove undesirable substances from the blood, passing them on to the bile and intestines for elimination from your body.

There are three built-in controls that increase or decrease blood flow and the delivery of nutrients and oxygen. Two of the controls are heart rate and stroke volume. Together they control cardiac output or the quantity of blood pumped per minute. The third control system is called resistance. It controls the volume of blood flowing to different regions of your body. Changes in resistance occur when tiny muscles surrounding blood vessels either contract, causing increased resistance and reduced blood flow, or relax allowing the diameter of the blood vessel to increase. The result is an increase of blood flow. These three factors: heart rate, stroke volume, and resistance work together to ensure that blood flow is maintained and controlled to provide oxygen, nutrition, and hormones to all tissues and cells, based on their needs. They also control the removal and transport of carbon dioxide back to the right side of your heart and on to your lungs to be exhaled. This control system responds immediately to changes in activity. When you exercise vigorously, blood flow increases to the active skeletal muscles and flow to the gastrointestinal system may decrease.

Your pump and transit system are affected by gravity and the free fall of space flight. When you are standing on Earth, blood flows down and tends to collect in your lower limbs. The reverse occurs when you lie down. Based on blood and fluid distribution there is a similarity between space flight and being recumbent, particularly if you lie with the head slightly lower than the feet. That is a condition known as head down tilt. It is often used in experiments on Earth to simulate cardiovascular and fluid shift changes found during space flight.

Recall the changes that were shown in Figure 7.6. They illustrate the fluid shifts and subsequent decrease in blood volume that occur during space flight. Along with the decrease in blood volume, your heart becomes physically smaller during a space mission. During the early hours of flight your heart may become larger due to fluid shifts, but this tends to change as you adapt to μg and the decrease in body fluids. On short missions, heart rate is slower in space than it was during preflight tests on Earth. A similar change is seen in laboratory rats while in space. There have been several instances of irregular heartbeats in astronauts during space missions, but most seem attributable to stress and are related to factors other than microgravity (Figure 7.7).

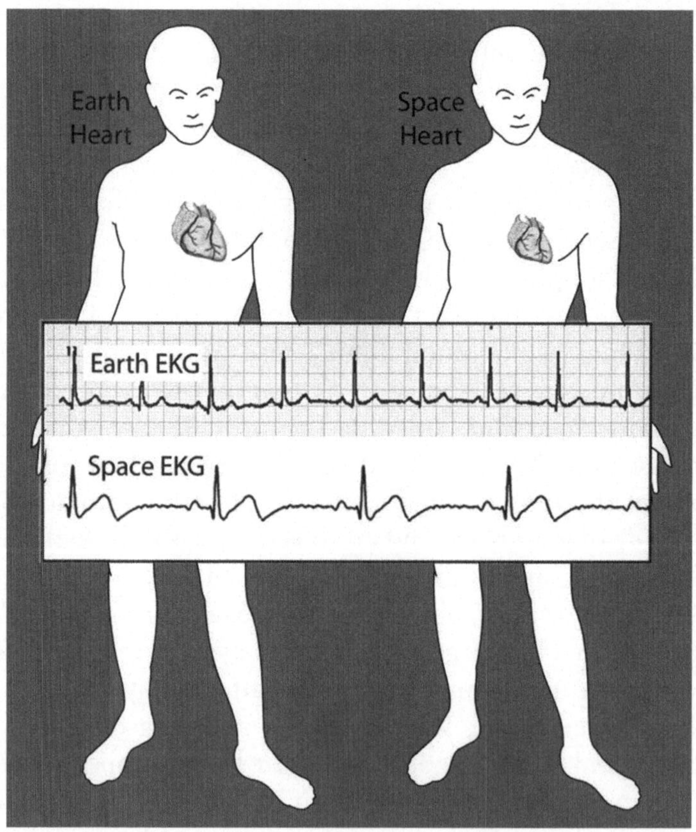

FIGURE 7.7 In space, the heart is smaller. It beats at a slower rate as shown in the EKG tracing in front of the figures. Without a gravity force to pump against, it is less work for your heart to distribute blood to all parts of your body.

The heart, the vascular system, and blood flow function well during μg exposure. What causes problems in the system is re-entry into the Earth's atmosphere. Deceleration takes the shuttle from a speed of 17,500 mph to only several hundred mph in order to land. The most abrupt decrease in speed happens when the space vehicle enters the atmosphere, about 60 miles above Earth. The rapid decrease in speed increases the gravitational force. At this time the shuttle is flying with the bottom of the shuttle facing forward and the atmosphere acting as a brake. Astronaut seats are positioned so that the gravity vector is from head to toe and blood is pulled down into their legs. Gravity increases two to three times Earth normal as the craft slows. The increased gravitational force causes excess pooling of blood in your lower limbs. This fluid shift, if not controlled, can cause the same decrease of blood flow to your brain seen in orthostatic intolerance; you may feel light-headed or faint. It is common for astronauts to wear inflatable pressure pants, often called a g-suit, to prevent excess fluid/blood movement into the lower limbs during the re-entry gravity increase.

Changes in the cardiovascular system during space flight are relatively minor. The heart beats slower and decreases in size. The only physically stressful times are during the increased gravitational force that occurs during launch and landing. Of these, landing is the greatest concern because at that time you will have a decrease in blood volume. Using pressure pants to prevent blood from pooling in your legs can help solve that problem.

Part C: Bones

But this man who thrust himself from the earth, who wore the stars of heaven in his hair, was guilty of overweening pride. In act most audacious, he had defied nothing less than the law of gravity. He was to pay dearly for such high imposture. The vertebrae, unused to their new columnar arrangement, slipped, buckled and wore out. Next, the arches of the feet fell. The hip joints ground to a halt.

Richard Selzer (1928–Present), from *Mortal Lessons, Notes on the Art of Surgery*

The fluid and the cardiovascular systems of your body are not the only parts that change in μg. The skeletal and the muscular systems work together as a single system that allows us to stand, walk, and run in our daily activities. Without a functional musculoskeletal system, we would be mere blobs of inert protoplasm on the Earth's surface, fixed in place.

We humans have 208 bones, connected together to form an interactive skeleton. The skeleton supports the rest of the body and gives us our shape. The bones of the legs and back are the parts that form the structure that allows us to stand and to overcome gravity.

Bone is a very dense material made out of minerals; particularly, calcium and phosphorus. The size and strength of bones are dependent upon the load that they normally carry. Thus, the bones in your legs are larger in diameter and stronger than the bones in your arms. They, along with the pelvis and vertebrae, provide stature. As we grow and mature, our bones gain size and strength. When we become more sedentary, bone mass is noticeably lost. The same changes happen in prolonged bed rest or hospitalization when your bones are no longer stressed. In addition many older people, particularly those who don't exercise, develop a condition known as osteoporosis in which their bones become more porous with larger internal canals and spaces. They lose strength and are less supportive. The bones of older individuals are particularly susceptible to breaking if a fall occurs. The pelvis is notorious in this regard. Bone is adaptable in response to the way and extent that it is used.

When you take your bones into space, they are only mildly stressed. Once again, you are in free fall in the μg environment. Physical stresses such as running and jumping which help maintain bones on Earth do not exist, and it is these physical maneuvers that help to maintain bones here on Earth. As a result, bones begin to adapt to their new environment of not being needed to withstand the physical stresses of daily living on Earth. Bone mass begins to be lost as it is not used, but it is not lost to the same extent in all parts of your body. Some bones, like the skull, ribs, and arms, are not really involved in defying gravity on Earth and are only very slowly changed by decreasing the effective gravitational field. It's a different matter if we consider supportive bones. Bone mass is lost most rapidly from your pelvis, then the tibia and femur in the legs, followed by your lumbar vertebrae (Figure 7.8).

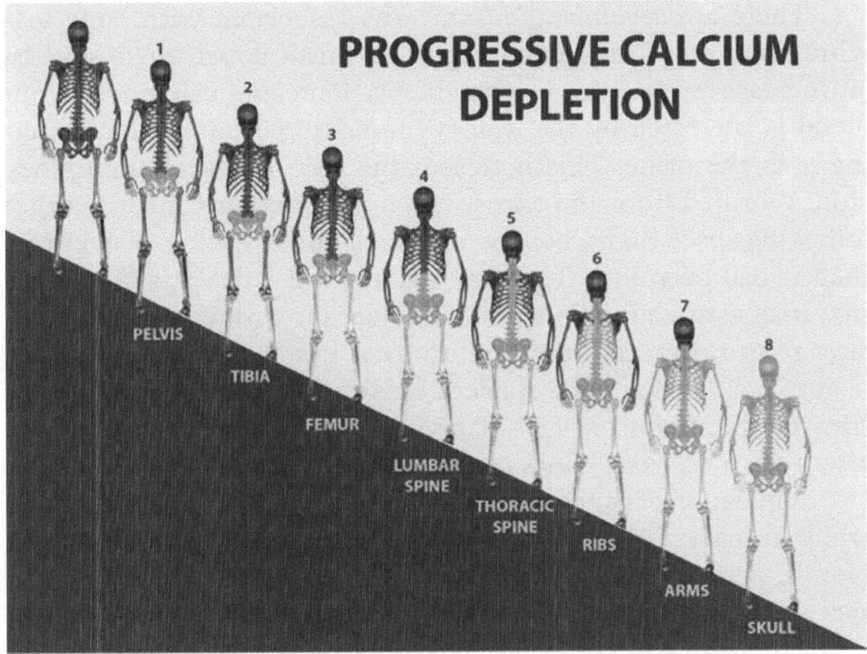

FIGURE 7.8 Although a great deal of bone or calcium is lost during space flight, it is not a uniform loss. The pelvis followed by the tibia and femur in the legs and the lumbar vertebrae lose bone mass most rapidly. Bony structures such as the skull and ribs, that do not support the body against gravity, lose very little bone mass or calcium.

The total quantity of bone mass lost during space flight is large and rapid. Medical doctors who are concerned with the health of astronauts consider bone loss as one of the most serious consequences of space flight. They are particularly worried about the quantity of bone that may be lost during a proposed 2.5 year exploration mission to Mars. The longer the space flight, the greater is the concern. At this time there is no indication that the rate of loss declines during long term exposure to μg. The Russians have provided the most information on bone loss during long space flights of over 300 days. The final answer isn't in. They have also found that bone loss is decreased by exercise that stresses the antigravity bones. It has been calculated that in space, bone is lost approximately 50–60 times as fast as it is in members of the population on Earth over 50 years of age. This is the group that is most susceptible to aging bone loss and osteoporosis.

There are several other concerns associated with bone loss while in space. When bone begins to break down, there will be an increase in calcium in your blood. Increased calcium in your blood is corrected by the kidneys removing calcium and excreting it in the urine. This increases the risk of developing kidney stones formed from the excess calcium. Another concern is that following space flight, bone is regained, but at a much slower rate than it had been lost. The factor is about 2.5 times. This means that if an astronaut is in space for 50 days it will require 125 days, after return, to regain the calcium that has been lost while in μg (Figure 7.9). This is not a concern for short space trips, but on long trips to the moon or Mars it could limit your physical activity after return till your bones regain "Earth strength."

A question of concern is how long will your bone loss continue while in space? Most space flights have only lasted several weeks. Now as new crews use the ISS, there are increasing numbers of spacefarers staying months, not weeks. Earlier, on the Russian Mir Space Station, several cosmonauts spent a year or more in space. At this time, there is no clear cut information that shows that the rate of bone loss begins to decrease in longer flights. This lack of

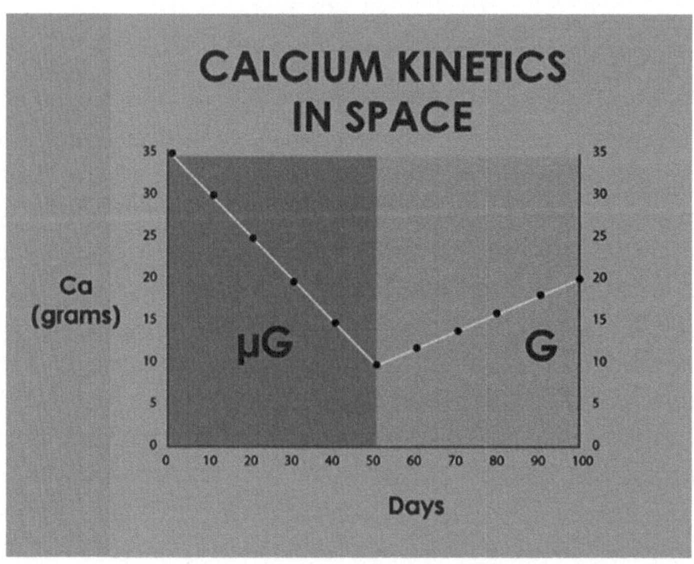

FIGURE 7.9 Calcium that has been lost during space flight is recovered slowly after return to Earth.

knowledge is certainly a risk factor in deciding the type of vehicle and ways to provide artificial gravity for longer journeys to Mars or expeditions on the surface of the moon.

Trips to the moon that are months in duration will be less hazardous to your bones than the same time spent in μg. The moon's small gravity field will presumably decrease the rate of calcium loss. However, it is likely that even with vigorous exercise while on the moon, it will be difficult to create stress factors similar to daily living on Earth. Bone will be lost. If you weigh 156 pounds on Earth, you will weigh only one-sixth as much or 26 pounds on the moon. The unaccustomed light weight will make it difficult to place the normal stress of Earth living on muscle and bone. Exercises that worked in staying fit on Earth may be of minimal value on the moon. You will need to have an effective exercise regime to slow both bone and muscle adaptation while on either the moon or Mars. That will make it easier for you to return to Earth. If you are determined to be a colonist and plan to make the moon your permanent home, there is no need to attempt to be Earth fit. You will just need to be fit to be on the moon.

A number of countermeasures have been developed with the intent of decreasing the rate of bone or calcium loss during space flight. The most popular and widely used is exercise. A large number of exercise devices have flown in space on US, Russian and International missions. There are several innovative pieces of exercise equipment on the ISS at this time, designed to provide a variety of exercise experiences. The goal is to exercise legs, arms, and the cardiovascular system in an attempt to block or decrease adaptation. Most of the equipment is modeled after exercise equipment in use on Earth; bicycles, treadmills, and resistance devices. On Skylab, with lots of open space in the laboratory, astronauts had the opportunity to devise unique exercise activities (see Video #001 Skylab Exercise in Appendix 1).

The exercise equipment on the ISS is designed to have a minimal impact upon the station's μg environment. The station was designed to be a floating laboratory providing a constant minimal gravity. Activities that might cause vibrations to the spacecraft are isolated by tethers or springs. They do not directly contact the surface of the station so that exercise induced vibrations are not transmitted to the spacecraft.

FIGURE 7.10 Treadmills are one of the most frequently used exercise counter-measures while in space. As they run, their leg bones receive impact as they do exercising on Earth. The treadmill turns because of foot and leg movement, there is no attached motor.

In the past, dating from Skylab in the 1970s through the Russian Soyuz and Mir Space Station, Space Shuttle, and ISS, exercise bicycles and treadmills have been extensively used as countermeasures. Theoretically, the treadmill is a better option than a bicycle because bungee cords are used to hold the exercising astronaut on the treadmill (Figure 7.10). The runner is not pushed up to the ceiling when running, but remains in position on the exercise device. Running provides impact forces on the leg bones and closely simulates running on a treadmill on Earth. The problem is that the body will be adapting all day, every day, to μg. Exercise will decrease the rate of adaptation for a few hours each day, but adaptation will still be occurring when not exercising. The final amount of mineral lost in susceptible bones will be the same with or without exercise. It may take years of exposure to decreased gravity to reach a stable endpoint in those portions of the skeleton that do not support weight. Non weight-bearing bones very slowly lose substance (Figure 7.10).

The ISS crew is testing another countermeasure while they orbit the Earth. The bisphosphonates chemical compounds of work well here on Earth to prevent or slow bone loss in elderly patients. This class of compounds is touted as likely to be of benefit in sustaining bone structure in astronauts. It would be an easy fix if bone loss could be diminished or even done away with by taking a pill or receiving an injection. The problem is, if you go into an environment where a particular function or structure is no longer needed your body will change. Drugs or chemical compounds are unlikely to be fully effective in preventing that.

Some preliminary planning for maintaining the bone density during a Mars mission is focused on creating artificial gravity using centrifuges. Some researchers feel that this can be accomplished with a Short Arm Centrifuge (SAC) housed inside the transit vehicle (Figure 7.11). Such a centrifuge would have a radius of

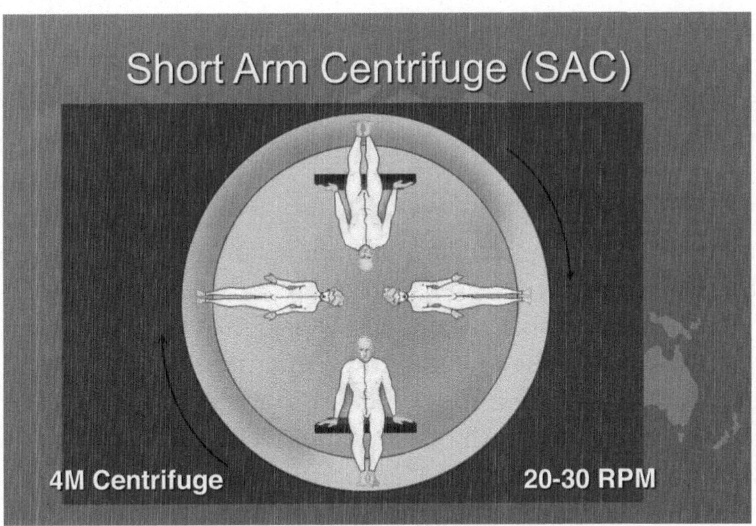

FIGURE 7.11 The short arm centrifuge has been tested on Earth as a means of reducing bone loss and cardiovascular changes in astronauts during long term flights in µg as required for a Mars mission.

about 2 m or 6 ft. It will probably need to be a bit bigger to allow for astronauts 6 ft or more in height.

The plan is to have the astronaut's head toward the center and their feet at the outside edge. When the centrifuge spins at slightly over 20 rpm the astronaut's feet will be subjected to a gravitational field similar to that on Earth. The gravity gradient will lessen towards the center so that only a small gravity field will be present at the head. The hope is that this device will inhibit bone loss during long term space missions. Trials of the device in space have not been conducted. It is not known how many hours per day might be required nor what the most effective diameter will be. Results of preliminary spinning tests on Earth indicate that many of the potential problems of rapid rotation on the neurovestibular system can be overcome. It is really amazing how easily we adapt to strange new environments. Perhaps sitting instead of standing on the centrifuge may help reduce adaptation in the pelvic area where bone loss is high (Figure 7.11). Having the hips at the outer rim of the centrifuge would increase their gravitational load while the centrifuge is running. Sitting will be of little or no benefit in maintaining bone mass in the legs.

As we gain experience on the effects of long-term lunar gravity exposure, another artificial gravity approach could be used for trips to Mars. The most likely solution, and one that is theoretically satisfactory, would be to utilize two space vehicles of equivalent mass, connected by a long tether or bridge of perhaps 100 m in length (Figure 7.12). One of the vehicles could be the crew quarters and contain food and other supplies needed for the voyage. The other would hold essential equipment that could be accessed by cable or bridge, as well as equipment and supplies for use on the Martian surface. Although more expensive than an onboard SAC, the tethered vehicle approach would provide a constant centrifugal force along the craft's outer rim where the crew would be living. Even better, the force could be varied by modifying the rate of revolution or the length of the tether. With a distance of 100 m between the two outer rims, the rotating vehicles would produce a force similar to Earth's normal gravity at 4.3 rpm and simulate Mars gravity at 2.6 rpm. Thus the astronauts could be adapted to the gravitational field of their destination on a 24/7 basis, not just

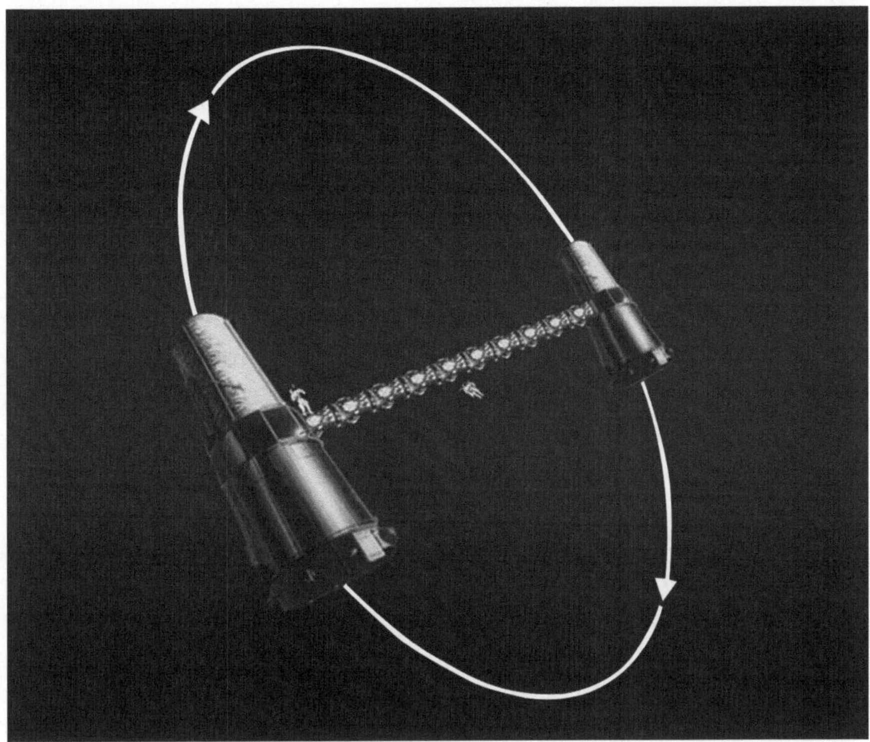

FIGURE 7.12 Having two vehicles of equal mass connected by a tether, or a tether/tunnel, can provide a gravitational field of any desired intensity along the outer rim of the vehicles.

a few hours/day. The desired gravitational force could be obtained at a slower rate of revolution with a longer tether or connecting bridge if there were any long-term unpleasant side effects of the rotation rate. The habitat capsule's interior would be designed so that the outer rim away from the tether attachment would be the floor of the living quarters.

NASA and other space agencies have minimal experience developing and deploying tethered vehicles at this time. Some preliminary experiments have been conducted and more will be designed as trips to Mars are being planned. This approach could be an effective solution to the serious bone loss anticipated during a Mars mission.

Part D: Moving Those Bones

> *Astronauts quite literally cannot generate the same amount of strength in their muscles. The mind of the Astronaut may remain sharp and alert, it's their muscle control-particularly the legs that are affected.*
>
> Doug Watt, Physician, Neurophysiologist

Bones provide the structure, but without muscles they are of little value. Understanding their relationship is simple. Each end of a muscle hooks onto a different bone across a joint. There are two or more muscles for each joint. At least one will be a flexor and one an extensor. These two muscle types can contract to bend or relax or extend the joint. For example, consider the muscles across the elbow. When the flexors contract, your elbow bends and you can lift a drink. When the extensors contract, the flexors automatically relax. They always work in concert. When the extensors contract, the elbow straightens and you can point at something in the distance. Continued action of many muscles working together run the great system that allows us to sit, stand, walk, run, and jump.

In your legs, antigravity muscles are your extensors. They keep your legs straight so that you can be erect. Their normal function is to allow you to stand, to move. Like a good servant, they get the job done without you thinking about it. Antigravity muscles are also called slow twitch muscles based on their speed of contraction. Their job is to provide a continuous low level of activity. This is the sort of thing you need to counter gravity. Other muscles are called fast twitch. They contract faster but tire more easily. You need them to jump, run, hit a ball, throw a ball or other sporadic physical activity that requires rapid action.

Those muscles that normally keep us erect lose their job when you're lying in bed at rest or floating in space. They quit contracting and just relax. During normal activity on Earth or any other gravitational field, they would be busy tending to the job of moving you about and defying gravity. In the μg environment of space, you are floating, not sitting or standing. Floating does not require muscle activity. Few of the contractions that are a part of everyday life in a gravitational field occur while in space. With muscles not being used, or sparingly used, they begin to adapt when their need is reduced. They waste away, atrophy, as antigravity muscles are not a requirement in free fall. It's the age old story; use it or lose it. Astronaut muscles waste away in spite of using exercise devices. The many space agencies dealing with human flight have developed unique and innovative exercise machines for space flight, designed to maintain muscle function. The trouble with exercise as a countermeasure to muscle loss is that you can't exercise 24/7, rest is equally important. If you did try to exercise continuously, you would become bored and quickly tire of the regimen. If you were able to be a full time jock, you could not exercise as well as conduct experiments, do spacecraft maintenance, eat, sleep, etc.

One example of the severity of functional muscle loss can be seen in the photograph of cosmonauts being carried in baskets following a mission to the Mir Space Station (Figure 7.13). After their prolonged months of exposure to a μg environment, they were not able to walk immediately after their return. Cosmonauts and astronauts, returning from long missions, can spend several months of time post-flight being rehabilitated to life in Earth's gravity field.

FIGURE 7.13 When cosmonauts returned from long visits to the Mir Space Station, their ability to stand and walk was compromised and they had to be carried off in recumbent couches. (Picture courtesy of NASA)

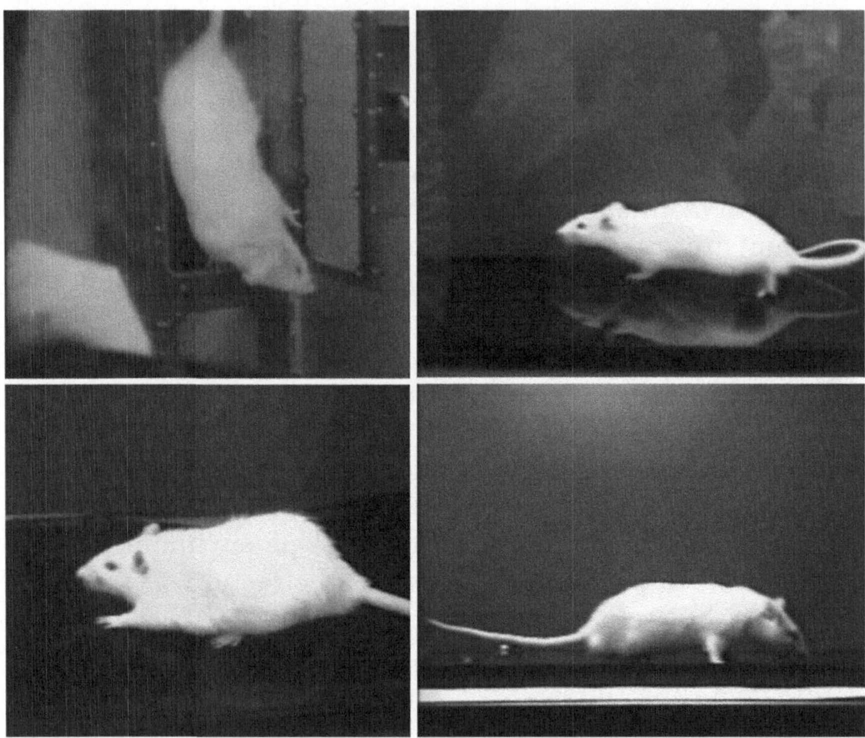

FIGURE 7.14 *Upper left*, a rat in space. He could move freely and was not at all limited in mobility. *Upper right*, an Earth based rat observed in a fish tank as a control. He also was mobile and walked freely. *Lower left*, a flight rat 2 h after return from a 9 day mission. He had severe muscle weakness and could not lift his body off the bottom of the tank. He was completely unable to fulfill his impulse to investigate by standing erect in a corner of the tank. He shuffled and crawled rather than walked. Pictured in the *lower right* is a flight animal 1 week after return. He had regained much of his mobility but still had some muscle weakness, carrying his hind limbs further forward under his body mass. (Video #002 Rat Behavior on Springer Extras shows the actual behavior depicted in these four pictures)

The speed of adaptation to space and loss of Earth gravity fitness is very rapid. The next series of pictures are frames from a video of a rat during spaceflight, 2 h and then 1 week post landing following a short 9 day space mission (Figure 7.14). The first picture is a rat in space in a work station freely moving over his cage. He appears to be well adapted to the new environment. One behavioral change is that he maintains a hold on one or more surfaces to prevent free floating. This reluctance to free float disappears after a longer adaptive period, particularly in a larger cage system. The second is a normal Earth control rat walking in an empty 10 gallon fish tank used as an observatory. The third is a space flight rat 2 h after landing, in the same tank, with its chest and abdomen both flat on the bottom or floor surface. The rat was unable to extend its legs and stand after only 9 days spent adapting to μg. Re-adaptation to Earth gravity for 1 week was enough to allow the rat to regain mobility in a gravitational field as shown in the fourth picture. Yet its gait was not entirely normal. Muscle weakness was still present. A few more weeks would be needed to become completely readapted to Earth life. It is not likely that the muscular systems of these young, growing rats as pictured in Figure 7.14 were completely adapted to space flight in only 9 days. Yet it would be difficult to imagine how the rat could become less mobile upon return. All of the leg muscles of space flight rats are decreased in size following return to Earth. The greatest decrease is in the extensor muscles that cause the legs to straighten.

Muscle weakness following space flight is largely due to atrophy of the antigravity muscles that allow us to stand, walk and jump. Examination of antigravity muscles of a space rat and a normal rat with a microscope shows that the muscles have become smaller (Figure 7.15). There is also a biochemical change that makes them less effective. They do re-adapt to Earth after return and recover strength as shown in the last section of the video. However, immediately upon return to Earth, space-induced changes are severe in rats and in people too. Neither are capable of physically performing in a normal manner. Recent results from a number of astronauts and cosmonauts, who spent a minimum of 6 months on the ISS, found that 70% of the individuals surveyed had muscle loss that was so great that it was unacceptable. These changes developed in spite of exercising countermeasures while in orbit.

Early in the Soviet human space flight program, a suit was designed to exercise antigravity muscles on a continual basis. It was given the name "penguin suit," because when it was first put on, you tended to waddle like a penguin as you moved (Figure 7.16). The suit has bands of elastic that require you to exercise your muscles just by wearing it. A number of cosmonauts who were in space for many months wore it for long periods each day in order to stay more fit.

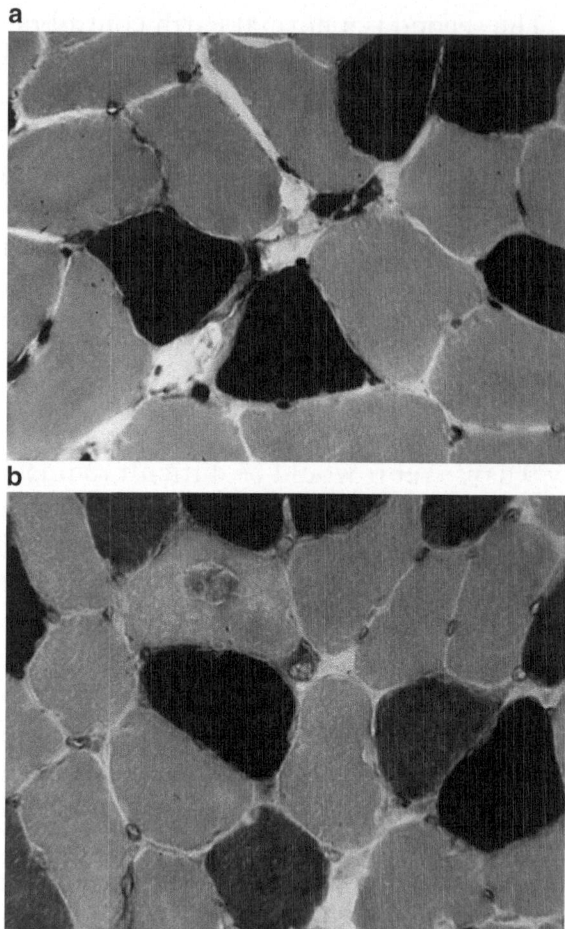

FIGURE 7.15 (**a**, **b**) Pictures of muscle fibers from the hind limb of two rats. The upper picture is muscle from a control rat on Earth. The lower picture is muscle from a rat following return from space. The diameter of the muscle fibers is larger in the control animal. The specimens were also stained with a chemical that detects muscle fiber type. The flight rat had an increase in what is known as fast twitch fibers, called so because they have a more rapid response. Slow twitch fibers that are not stained darkly are associated with serving an antigravity function. (Picture courtesy of D. Riley)

FIGURE 7.16 Drawing of a cosmonaut wearing the penguin suit designed to force his antigravity muscles to contract even if he was floating in ug.

Because of problems with mobility, landing on Mars after a 6 months exposure to microgravity is probably not the best solution. A better approach could be to design the transit vehicle to have a gravitational field. If large enough, with the right configuration, it could provide a gravitational field just right for the destination. One solution that is under consideration would be to have two space vehicles, each with the same mass, connected by a tether system as shown in Figure 7.13. In the tether design, the direction up is towards the opposite capsule. If the astronauts are pre-adapted to the Mars gravitational field while in transit, they can set up or activate their living facility and commence exploration as soon as they arrive without devoting a period of time to adapting to the Martian 0.385 gravitational field. They could also spend the 6 months return voyage re-adapting to Earth's gravitational field.

Part E: Up Is Relative

One of the questions they asked us on our first flight was, close your eyes now, how do you determine up? With his eyes closed he couldn't tell. Up and down had vanished.

Robert Parker (1936–Present), Astronaut

All of your body's systems; blood, bone, heart, muscle, etc. need to work together like an assembly line in a factory or a well trained sports team. Teamwork between parts of the body is necessary if you are going to live up to your potential. The job of organizing and running your body is the responsibility of your control system, better known as the nervous system. In many ways it is like having a built-in personal computer. Portions of the computer can function by themselves without conscious input, while other functions need direction. You don't have to think about breathing or telling your heart to beat faster or slower. Those activities are hard wired. Deciding to run a race, asking or answering a question are under your voluntary control.

It takes some years to get your whole body system organized and functioning. Even then, the circuits get challenged and overloaded by things like puberty and hormones. The nervous system comes with most of the required hardware and you add software programs as you age. You probably know it as going to school and maturing. The system runs on a combination of electrical circuits and chemicals released at junctions between nerves. The chemicals either speed up or slow down body functions. Sometimes they turn systems on or off. Electrochemical conduits, or nerves, carry control commands from the Central Nervous System (CNS), your brain and spinal column, to the rest of your body. The CNS, in turn, receives electrochemical information about your body and its environment from a variety of sensory receptors. Some of that information reaches our conscious minds, but much of it is handled, integrated, and acted upon without any awareness on our part. There is no reason for you to be bothered by superfluous information. To walk or run, our leg muscles work together to make us move. They contract and relax without the operator, you, being aware of all the commands necessary to make it happen.

Information about your environment and the state of your body enters the nervous system through the senses. Five of the senses are well known; seeing, hearing, smelling, tasting, and touch provide information on temperature pain, position of body parts, and pressure. Collectively, the information about body position, touch, and pressure is called 'proprioception.' On Earth, it is not necessary to continually check on your arms and legs, as the nervous system in the muscles provide all that information to your brain. However, these five sensory systems don't tell the whole story.

There is a sixth sense located in the inner ear called the vestibular apparatus (Figure 7.17). The vestibular apparatus is just a small part of our nervous system. It is directly related to our

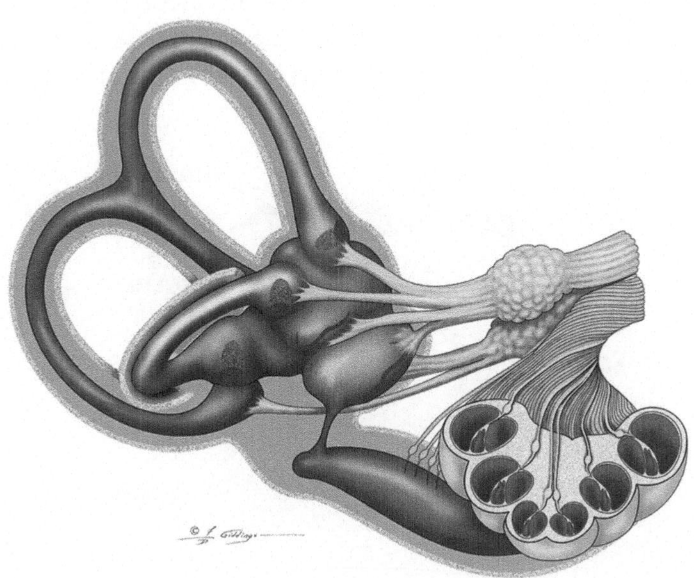

FIGURE 7.17 Vestibular apparatus. The sensory receptors are in the ampulla at the ends of the semicircular canals. The utricle and saccule detect position with respect to gravity. The yellow nerve carries sensory information to the brain.

perception of position, equilibrium, balance, and acceleration. It provides information about position of your head with respect to gravity. It also detects acceleration: both linear, forward or reverse, side to side, rotary like a twirling skater or tumbling like an acrobat. It does not detect velocity or speed, but it does detect changes in velocity. Faster, slower and the direction of the gravitational field complement vision and touch/proprioception in helping us be aware of where we are and what is going on in the environment around us.

The basic mechanism of detection in the vestibular apparatus is the bending of tiny hairs called cilia on specialized receptor cells that are located in the inner ear. Your body uses two different tricks to bend these tiny hairs. First the hair cells in the utricle and saccule line the walls with the cilia projecting inward as (Figure 7.18). They are imbedded in a gelatinous mass containing many dense crystals of calcium carbonate. The crystals are called otoliths, which literally means ear stones. Being heavier than the gel, the otoliths move the gel in the direction of the gravity field when your head's orientation is changed. This movement causes the cilia to bend. The bending sends an electrochemical message to the brain, a nerve impulse. The system is very sensitive when your head is in a perfectly vertical alignment with gravity.

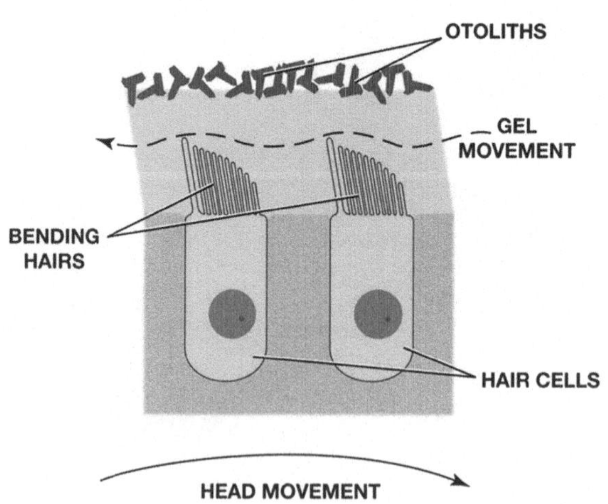

Figure 7.18 Otoliths in a semi-fluid gelatinous mass containing sensitive hair cells (cilia) provide a signal to your brain regarding the position of the head and linear acceleration.

For instance, a change in head orientation as small as 1° can be detected. The sensory portion of the utricle and saccule also detect linear increases or decreases in speed. A quick start to run forward produces a stimulus similar to starting a somersault.

Semicircular canals use a different system to cause cilia to move and detect head movement. The canals are filled with fluid that is freely movable. At one end of each of the canals there is an enlarged area called an ampulla (Figure 7.19). Hair cells and their cilia line the inner surface of the ampulla. Each of the three canals is in a different orientation so that any change in head rotary movement will be detected by the hairs in one or more of the ampullae (Figure 7.19). The canals and ampullae function in a slightly different way than the utricle and saccule with their gel and tiny stones. When you turn your head, the canals move since they are encased in your skull. The fluid tends to move more slowly in the opposite direction causing the hair cells projecting into the fluid to bend, initiating a signal to your brain. Several head movements cause stimulation in more than one canal. The three basic movements are called pitch, as when you pitch forward doing a somersault or nod your head 'yes' in an up and down motion. The second motion is called roll. It is a movement like a cartwheel or

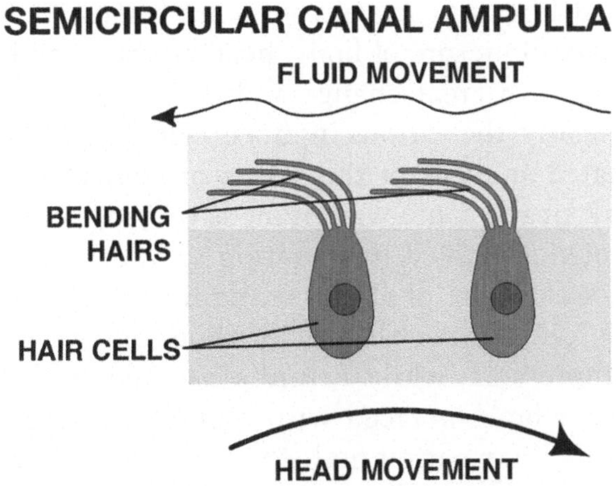

SEMICIRCULAR CANAL AMPULLA

FLUID MOVEMENT

BENDING HAIRS

HAIR CELLS

HEAD MOVEMENT

FIGURE 7.19 The semicircular canals are filled with fluid. When your head is rotated in one or more planes, the fluid in the canals lags the head movement. Since the hair cells are indirectly fixed to the head, they bend in the opposite direction. This initiates a signal to your brain.

bending your neck to touch your ear to your shoulder. The third movement is called yaw and is a type of movement executed by a twirling skater or ballet dancer. Shaking your head to signify 'no' activates the same portion of the system. The semicircular canals do not detect motion itself, but instead a change in motion, acceleration or deceleration.

If you sit in a swivel chair with your eyes closed and the chair spins at a constant speed for several minutes, you will notice the initial acceleration. You lose that moving sensation when the fluid speed in the canal matches the speed of the body. If the chair suddenly stops, you will feel that you are rotating in the opposite direction as the fluid will keep moving for a brief period. It is easy to see how this system works. Use a glass of water with a single ice cube floating near the edge. Rotate the glass. The ice cube and water movement will lag behind the movement of the glass. If you continue to rotate the glass at the same speed, fluid and ice movement will catch up to the speed of the glass. Water, ice, and glass will be rotating at the same speed. With a sudden stop of the glass, the ice cube and water will briefly continue to move.

Vestibular apparatus function is linked with both vision and proprioception, the sense of touch/pressure and position. These three sensory systems have worked together as partners for millions of years in Earth's gravitational field. Sensory information is transmitted to different parts of your brain where it is integrated to provide an overall picture of body, head position, and movement. Exposure to μg is a great change and challenge to these sensory systems. Basically, the various inputs no longer present a coherent and coordinated analysis of the status of your body. Some components work in the same way in μg as they do on Earth. Others send erroneous or garbled information to your brain. An overall description of the state of affairs while in space is "sensory conflict" (Figure 7.20). One sense provides one analysis, another sense a different impression, while a third sense may provide no coherent input. Your brain is faced with a problem. What to believe? The conflicting information no longer provides a correct and unified analysis of the state of your body.

What really happens during space flight to your sensory receptors? The gravity receptors in your inner ear do not provide any information as there is no weight to the otoliths. The cilia on the

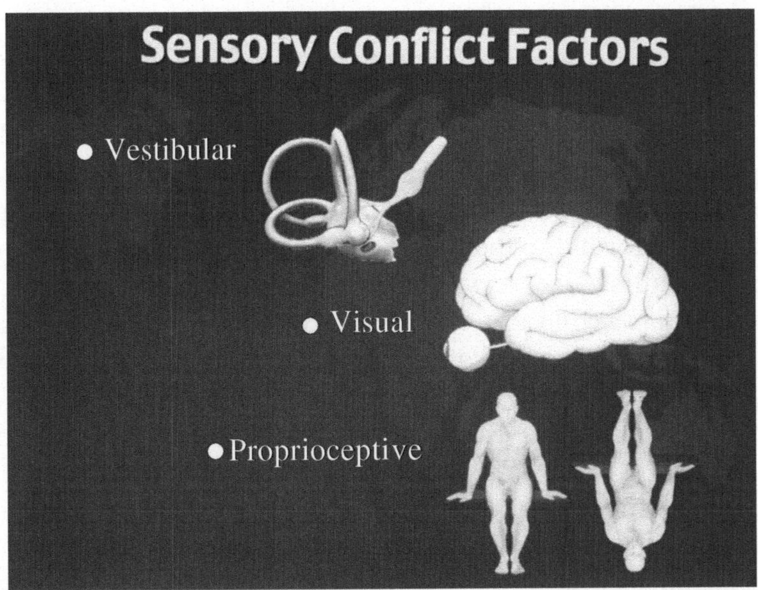

FIGURE 7.20 Conflicting information is provided to your brain by the various sensory systems. In this diagram, the gravity receptors in the vestibular system may not really be activated if there is no change in motion. Your eyes and proprioceptive systems may add to the conflict by misinterpreting body position compared to orientation in the space vehicle.

hair cells don't bend regardless of your position, except when you accelerate. While in space, your limbs float freely when relaxed and may be in any position. Holding an arm straight out or over your head is an effortless action. If floating, the sense of touch on your skin does not receive constant pressure contact. Vision is still functioning, but the impact of seeing crew members floating in strange positions, upside down or backwards, can be disturbing. A colleague reported after his shuttle mission that he had been doing very well with minimal problems and had started to eat an apple. Just at that time, one of his crewmates floated by. He was upside down and moving backwards. There were too many conflicting inputs. My friend began to feel upset, discarded the apple, and got out a 'barf' bag.

In μg, if you are sitting on a surface with the aid of handholds to get firm contact between your buttocks and the surface, your brain will interpret that direction as down regardless of visual information to the contrary. You could be sitting on the ceiling, but with that proprioceptive input, you will feel that the ceiling is down. For millions of years, we have known that if we sit on something and have detectable pressure on the buttocks, that direction is down.

While you are in space the semicircular canals will still be able to process rotary acceleration. The utricle and saccule will detect linear acceleration. These familiar inputs may not coincide with other information being processed by your brain. This jumble of input results in sensory conflict. The result of this conflicting information is a condition that was earlier known as Space Adaptation Syndrome (SAS), back in the days when it was inappropriate to suggest that astronauts got sick (Figure 7.21). Today, it is more routinely known as Space Motion Sickness (SMS) as there are many similarities between this discomfort, while in space, and motion sickness on a boat, an automobile, or some amusement park rides.

SMS affects more than half of first time space flyers. The severity ranges from mild discomfort to vomiting. In most cases, it is over in several days as your brain adapts to the unusual barrage of sensory input. However, some people will continue to be plagued just as sea sickness bothers some passengers for the entire trip on long ocean voyages.

There is a solution to the problem of SMS. The drug, promethazine (phenergan), has been effective in eliminating symptoms of

FIGURE 7.21 Space Adaptation Syndrome or more accurately Space Motion Sickness is believed to be due to sensory conflict in the nervous system.

SMS in astronauts but it is a depressant and causes drowsiness. Space Shuttle flights are tightly scheduled with little free time. Drowsiness and depression would tend to inhibit successful completion of planned activities, but then so does motion sickness.

It is helpful to not make unnecessary head movements if suffering from SMS, as head movements tend to increase symptoms. While watching videos of astronauts in space, it is relatively easy to spot those with SMS. They tend not to turn their heads rapidly, but instead slowly rotate their bodies when they need to look in another direction.

Visual inputs can be deceiving, but are not specifically related to SMS. In humans, the portion of the brain that receives and processes visual information is much larger than the processing centers for other sensory inputs. Visual interpretations can sometimes override other information. In the ISS all light comes from one side. In a normal gravitation field on Earth, the light source is interpreted as the ceiling. Having lights only on one side provides an inferred orientation of up. Yet, operational equipment is located on all four sides (Figure 7.22). Note in the artist's picture,

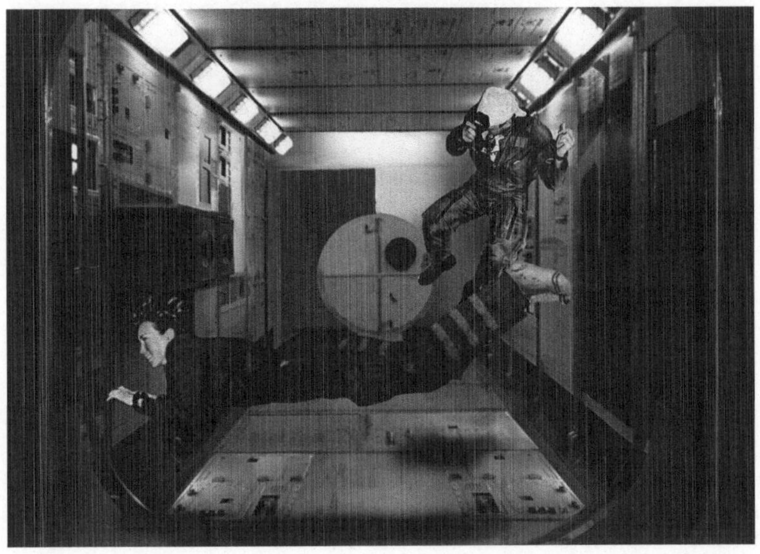

FIGURE 7.22 An artist's concept of working conditions on the ISS. Equipment is located on all four surfaces, yet light only comes from one direction, at the *top* in this picture.

the lady astronaut in the foreground casts a shadow on the lower level as she floats.

According to Joe Kerwin on Skylab, "All one has to do is rotate one's body to a new orientation and whammo! What one thinks is up is up." After the mission he said,"Up is over your head, down is below your feet." While in space, whether a wall is a floor, ceiling, or wall depends on you. It can be anything you want it to be.

Now consider the image in Figure 7.23. The shaded portion in the bottom of the left circle makes the circle look convex. Our Earth adapted vision processing "knows" that light comes from the sun or in a room comes from above. Conversely, in the right hand circle with shading on the top half, we perceive the circle as concave. Now rotate the book slowly to the right or left. Your vision processing centers will change the perception of convex/concave on the two circles at about the half-way mark. This can also be seen on the DVD where the circle will slowly rotate.

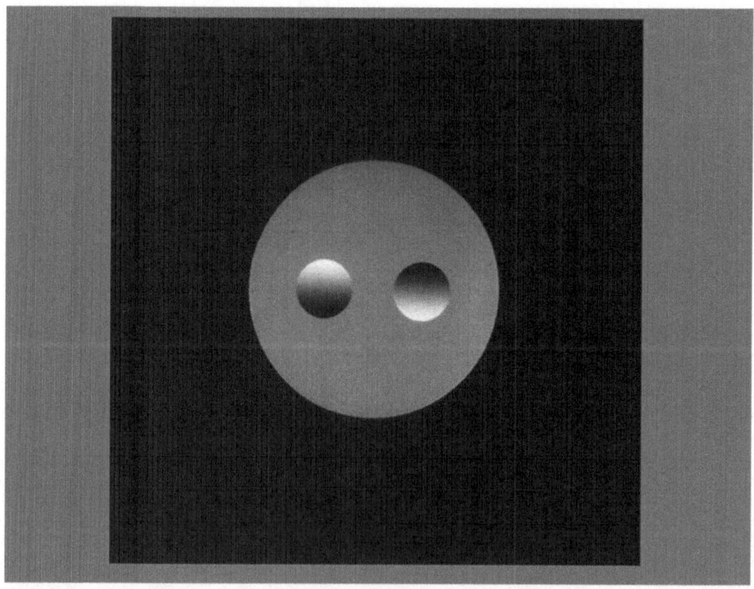

FIGURE 7.23 The circle with the shading at the *top* appears concave and when shaded at the *bottom*, convex. (Picture courtesy of J. Buckey. See Video #003, on Springer Extras to watch the change from concave to convex as the circle rotates)

Part F: Your Personal Castle

> *Add to this the biological improbability that makes each member of our own species unique. Everyone is one in 3 billion at the moment, which describes the odds. Each of us is a self contained, free-standing individual, labeled by specific protein configurations at the surface of cells, identifiable by whorls of fingertip skin, maybe even by special medleys of fragrance. You'd think we'd never stop dancing.*
>
> Lewis Thomas (1913–1993), from *Lives of a Cell*

Imagine your body as a castle that is protected inside and out by your immune system. Like any good protection system, there are several lines of defense. The first defense of a well constructed castle is the moat. The moat provides a physical and chemical barrier between an invader and the castle wall, your skin. With respect to your body, various secretions such as mucus in your respiratory tract, tears, and saliva contain chemicals that help protect you as the water in the castle's moat protects the castle. In addition, skin provides a relatively impermeable protective layer like the walls of a castle.

What if the moat and castle walls are breached by invading organisms? Inside the castle, your body, there are a number of defender cells organized into two highly coordinated fighting forces. The first defender cells that an invader would contact, function as "attack dogs." These cells are called macrophages, which literally means 'big eaters.' In Figure 7.24, macrophages are represented by Mac, the attack dog. Macrophages swarm to the place where the castle wall has been breached, then they identify and devour the enemy. Meanwhile, they signal two armies of fellow defenders that an attack is occurring. In Figure 7.21, they are depicted as soldiers. These soldiers are continually roaming the castle ready to repel invaders.

The magic of the immune system is the way that it detects invaders. It is a simple concept, but wonderfully complex in how it is carried out. The membranes surrounding every cell in your body contain proteins that are unique to each individual, like chemical fingerprints. Just as fingerprints are different and recognizable for every person on Earth, cell membranes are different and recognizable to your body's guard cells. The immune system's macrophages and "soldier" cells recognize these unique proteins

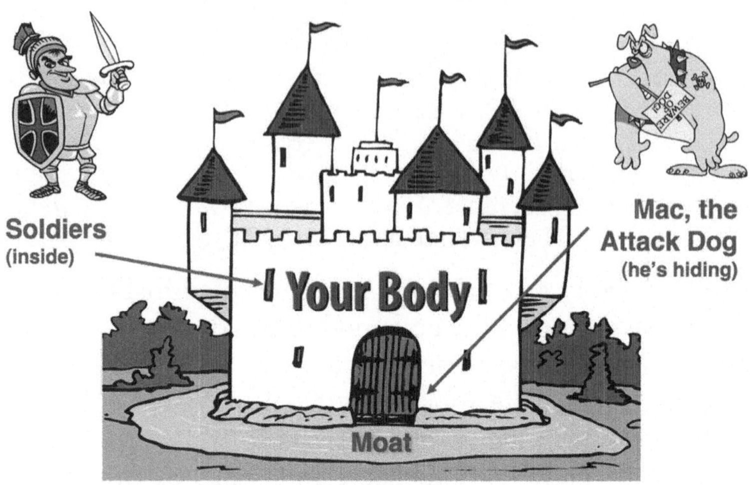

FIGURE 7.24 Your body is a castle and like any castle has multiple protection strategies. (Picture courtesy of G. Coulter)

and determine the difference between self and non-self, or foreign invaders. That ability to detect self from non-self is the real basis for immune system function. Without that ability, you are lost with no way to protect yourself from disease agents.

When a foreign entity is identified in the body, the immune system will work to eliminate it by a number of processes. Eating the foreigner by phagocytosis (cell eating) is "Mac the Attack Dog's" method. The two armies the macrophages put on alert each have a unique method in dealing with invaders. One army engages in cell to cell combat and injects strong chemicals into foreign cells causing their inner workings to break down. Ultimately they disintegrate. This is called 'cell-mediated immunity.' The second army uses a different tactic. They neutralize invaders with what could be termed "chemical smart bombs" or antibodies that attach specifically to the foreigner and make it inactive. The action of this second army of cells is called 'antibody-mediated immunity.' Immune cells throughout the body are continually asking the question of everything that they contact, are you friend or enemy, self or non-self? They then act accordingly to keep your body safe.

How and why does the immune system change during space flight? The how, at least in short missions, is easy to answer but the why is more perplexing. Every other system of the body that we have considered changes and adapts to be more effective in the new environment of space. One part of the immune system, cell-mediated immunity, partially shuts down becoming less effective. This change has been found in both humans and laboratory animals. It is definitely not a beneficial adaptive response.

Physiologically, some of the changes in other body systems that occur in μg could have an effect on immune system function. One plausible explanation is that the immune system changes in space are similar to some of the immune system changes that have been identified with psychosocial stress or physical and emotional overload on Earth. The ability to respond to foreign invaders and mount an effective defense is reduced. If this simple explanation is the answer, when space travel becomes more routine, stresses associated with the journey into μg should decrease and immune system function should be more normal. In any case, currently known changes to the immune system during space flight are of concern. No one wants to leave on a long voyage and be more susceptible to disease. Radiation can also damage the immune system. We know that voyagers away from LEO will face increased radiation exposure, but the long-term effects on the immune system remain unknown.

Fortunately, there haven't been any cases of disease in astronauts that can be traced to immune system inhibition. Much of the information collected so far has been from short missions that have had very intensive operating schedules with little time for relaxation. At the other extreme is the yet to be scheduled operational agenda for a 6 months Earth to Mars flight, a stay on the surface of 1.5 years and 6 months voyage back. Boredom is more likely a concern with excess spare time during the transit to and from Mars. The trip to Mars would be followed by a very busy schedule that would involve setting up living quarters and exploring a portion of the planet. Isolation, small crew size, and the long period of commitment could result in chronic psychosocial stress that might inhibit immune function more profoundly.

Astronauts on Mars will also have to learn to deal with time lag in conversations with Earth. It will be a new, troublesome experience.

The length of time from asking a question to receiving an answer will vary from almost 10 min when the Earth and Mars are close together, to over 40 min when the two planets are far apart on opposite sides of the sun. Carrying on conversations with even a 10 min lag time between question and answer will be just an unusual circumstance that could cause stress.

Sensory isolation from home will be a different stress when you are millions of miles away from your family, instead of 200 miles up in Low Earth Orbit. The bottom line seems to be that flying in space has the similar effect on the immune system that chronic stress does, especially its cell-mediated branch. As technology advances and space travel becomes more common, generally accepted and familiar to the public, some of the stress-related symptoms seen today should lessen, or be eliminated.

8. Behavior in Space

We do not stop playing because we grow old
We grow old because we stop playing.
Benjamin Franklin (1706–1790), A Founding Father of the United States, Statesman, Scientist
Getting old is a fascinating thing. The older you get, the older you want to get.

F. Scott Fitzgerald (1896–1940), Author

Part A: Sag, Shuffle, Fall

Gravity is the bane of our elderly years. As we age, we become less physically fit. Our sense of balance and equilibrium are less effective and our muscular reflexes become slower. We sag. We lump. We shuffle. We fall. Growing old is not for the faint hearted. It requires hard work and perseverance. The force of gravity complicates the problem. Since Earth's gravitational force causes problems for seniors they might consider moving to a lesser gravity field.

From a lifetime perspective, from the moment of conception, you spend the first 9 months effortlessly floating in Mother's womb, relatively safe from falling, accidents, and injuries. Why not end life in an equally non-stressed and safe environment? In the not too distant future, you could consider spending your retirement years in space. Admittedly, today, given our relatively unsophisticated launch methods, there is a brief period of hyper-gravity that may be uncomfortable, but new, less traumatic launch technologies are in the works. The race is on to make it possible for the public to easily access space. Individuals will go to space, including the moon, for exploration, science, tourism, adventure, and to live comfortably during their later years. They will be able to abandon, or at least minimize, the struggle against gravity. That battle is a constant part of daily life on Earth.

R.W. Phillips, *Grappling with Gravity: How Will Life Adapt to Living in Space?*, Astronomers' Universe, DOI 10.1007/978-1-4419-6899-9_8,

Think for a moment about the type of physical surroundings that would make your golden years more pleasant. Now think about the physiological changes that occur in space flight. The closeness of the match is obvious. The feeling of weightlessness means that there is no need to climb stairs as there are none. Any object, regardless of its weight on Earth, floats freely and can be easily moved about with little or no effort. Your heart, while in space, gets smaller and beats slower as there is no gravitational force to pump against. If you have a weakened or damaged heart that is partially functional on Earth, you might do perfectly well in an environment that requires a less active blood circulation. Broken hips or cerebral hemorrhages from a loss of balance and falling wouldn't be a concern. You can't fall to the floor or down the stairs in μg. On the moon with its minimal gravity, falling will be a slow process, much less traumatic than a fall on Earth.

In space once acclimated to μg, you will find it easy to move around the living quarters with a minimum of effort. Creaky and painful joints won't be stressed. There will be no need for wheel chairs; you can leave them on Earth. With arm mobility, everyone freely floats from one location to another. Bed sores will not occur because pressure points will not be stressed by the body's weight. If you have trouble with varicose veins, they will disappear due to upward fluid shifts as soon as you reach orbit (Figure 8.1). All in all, living in μg is not physically demanding in any way.

The most beneficial environment for older bones and body parts would be μg. The cost of building and maintaining large complex space structures is incredibly big and it may be better to consider a second option. That would be life on the moon, with a gravity force only one-sixth as strong as Earth's. The moon, as a tourist destination, is a component of the space program in several countries as well as a number of commercial entities.

It is easy to visualize a retirement/senior care complex operating in conjunction with a tourist vacation spa. Many of their requirements are similar. The two facilities could share components, thus decreasing the need for completely redundant systems.

The weight and complexity of space suits at this time makes it unlikely that most senior citizens could take advantage of strolling on the moon's surface. Instead, they could be passengers in

FIGURE 8.1 After a lunar colony is firmly established with adequate pressurized space, senior citizen living facilities may well become available. With a minimal gravity field, falling will be a minor hazard and varicose veins will disappear. It should be easier to stand erect if you weigh so much less than you do on Earth.

pressurized moon rovers with the opportunity to see new sights while riding in a comfortable vehicle.

Some might rightly ask, can elderly individuals with medical problems go into space? Career astronauts are, for the most part, in the prime of their lives and are required to maintain their physical fitness. Frankly, we don't know what infirmities would be limiting at this time. The one well-documented case of a senior citizen with potentially serious medical limitations who went to space as a tourist is very promising. At first he was turned down for a trip on the Russian Soyuz vehicle to the ISS because he had serious problems with both lungs and heart. After rigorous testing in the United States and Russia, he was cleared for flight and had no medical problems during launch, landing, or the 10 day stay in μg.

As private companies develop the capacity for space tourism, individuals with varied medical backgrounds will have the opportunity to fly in space. As this occurs, more complete information will become available as to what medical conditions can tolerate space flight and what may be grounds for staying at home. During this process, undoubtedly some errors will be made, but groundbreaking pioneers have always known that they are in essence "leading the pack." They must be prepared to face the potential consequences of "being the first." Like anything in life, there are pros and cons. The negative aspect of moving to a lunar gravity environment is that the rate of bone loss would likely be increased making it impossible to return to Earth. But then, for most elderly patients the move to a nursing home means never going back to managing a household, driving a car, or mowing a lawn.

If the thought of leaving family and friends behind forever seems difficult, reflect on the changes in communication that had started to become commonplace in the waning years of the twentieth century. With only a few years into the twenty-first century, video phones and video computers have appeared and are replacing standard cell phones and computers as a means of keeping in touch. Given the rapidly changing communication options, it can be safely predicted that direct face-to-face electronic communication with your family back on Earth will be easy from a lunar site.

The cost of space flight at this time is extraordinarily high and lunar elderly care would be prohibitively expensive for all but a few. However predictions for the future are very encouraging. The cost of reaching LEO will dramatically decrease in the coming years and access to space won't be limited to a chosen few. The moon is only a couple of days away using current technology and will become an easily reached destination for many in the foreseeable future.

Part B: Loops and Somersaults

> *The truth is, the science of Nature has been already too long made only a work of the brain and the fancy: It is now high time that we should return to the plainness and soundness of observations on material things.*
>
> Robert Hooke (1635–1703), Natural Philosopher, from *Micrographia* (1665)

In microgravity, an unanticipated, strange, perhaps bizarre behavioral pattern occurs in animals that fly or swim on Earth. They turn in loops, somersaults, and spirals. So far, we have had limited

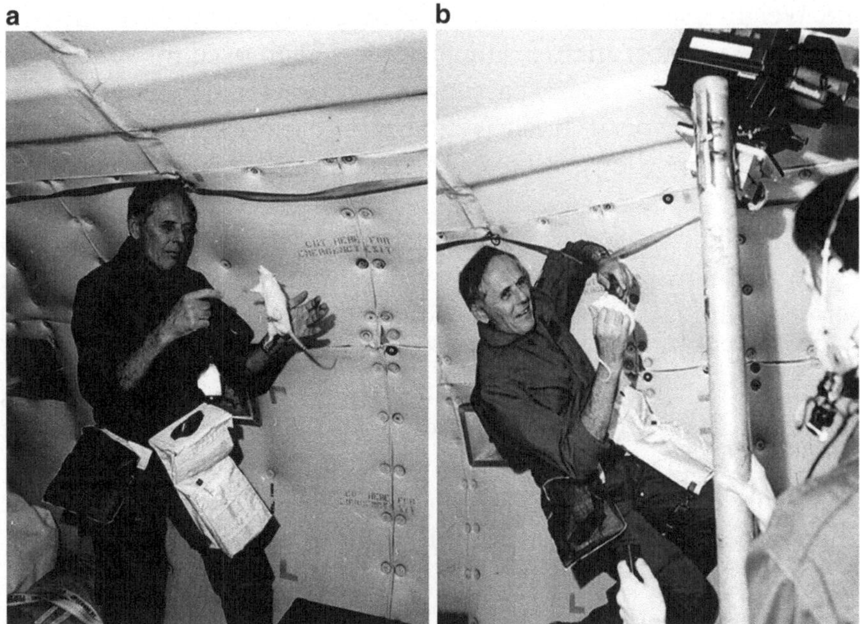

FIGURE 8.2 (**a**, **b**) Rats and other mammals as well as frogs extend their limbs when in μg. This picture was taken on the NASA KC 135 airplane that creates brief periods of μg by flying in parabolas. The author is accompanying the floating rat with outstretched legs on the left. On the right, the rat is holding on with all four legs and feet and even has his tail wrapped around the author's wrist.

opportunities to observe the behavior of flying and swimming species with sufficient room to rotate in their habitats, yet the pattern is consistent. Jellyfish, fish, tadpoles, and birds turn loops or somersaults when they are first exposed to μg. Nothing like this is seen in ground based animals during parabolic flights or while in space.

Rats, prairie dogs, and lizards search for something to hold on to by extending all four limbs, reaching for a stable point. Adult frogs do the same even though tadpoles loop. Figure 8.2 shows a rat with all legs extended during a period of μg on the NASA plane that flies out of Johnson Space Center in Houston. On these flights, the airplane flies over the Gulf of Mexico and performs a series of parabolas, usually 40. ach parabola creates about 20–25 seconds of free fall within the airplane, briefly simulating the μg of Space. The airplane is called the KC135, or more colloquially the Weightless Wonder or the Vomit Comet. This last name is probably the most fitting. The plane has been used for so long for these flights and so many passengers have gotten ill, that it has the stench of stale vomit. It is easy to begin to feel queasy just getting on board before they even start the engines. Still it is a great trip, I was always ready to go again.

The flights are used in training to give experience in μg or to test equipment to ensure that it will operate in a ug environment before sending it to space. The parabolic flight path of the KC135 is shown in Figure 8.3. For each period of μg there are two periods of hypergravity. As can be seen from Figure 8.3, the 25 seconds long drop in altitude on the downward portion of the parabola is about 8,000 ft, roughly a mile and a half. It is undoubtedly the most fantastic rollercoaster type ride ever devised. The greatest land base rollercoaster can't compare to a ride on the Weightless Wonder.

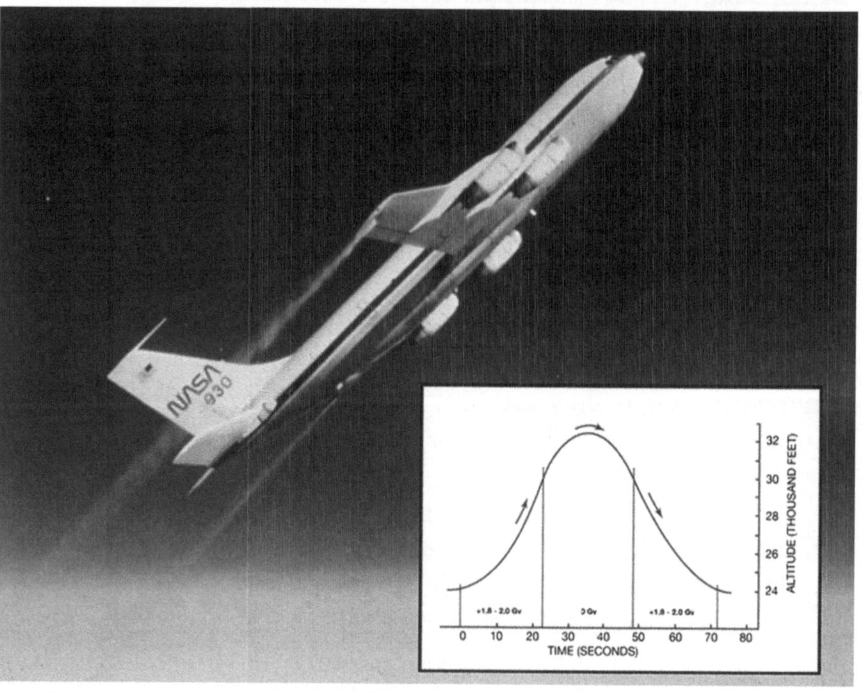

FIGURE 8.3 The KC-135 airplane flies a series, usually 40, of parabolic loops over the Gulf of Mexico near Houston. Each parabola provides periods of hyper-gravity as well as 20–25 seconds of μg. It is a great trip for roller coaster enthusiasts. (Photo courtesy of NASA)

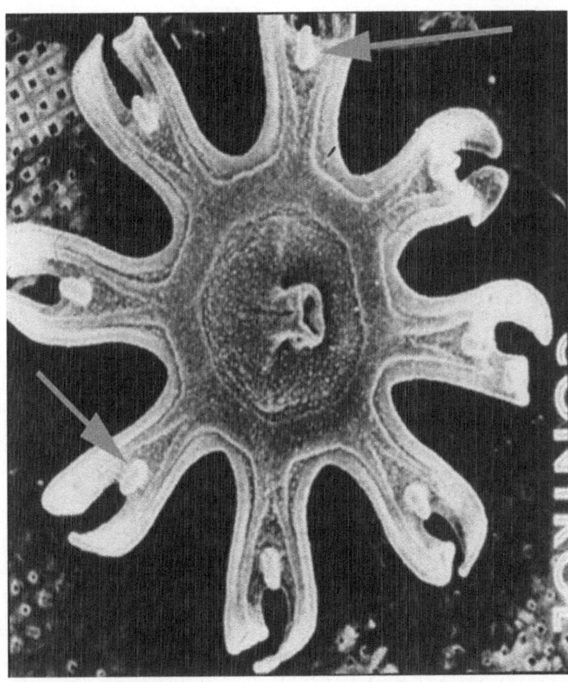

FIGURE 8.4 Jellyfish ephyrae. Scanning electron micrograph of the underside of an immature free swimming stage of a jellyfish. When swimming all eight arms contract simultaneously, causing the animal to move up. Gravity receptors are at the tips of the *arrows*. During space flight these animals tended to swim or pulse erratically followed by periods of inactivity. When swimming in space, they often performed loops. (Picture courtesy of D. Spangenberg)

The simplest, multi-cellular swimming organisms to travel on the Space Shuttle were very young and immature jellyfish. These little creatures not only have gravity receptors, but the receptors are clearly visible on the animal's lower surface. Figure 8.4 is a scanning electron micrograph of an immature free swimming stage of the jellyfish Aurelia. It is called an ephyrae. At the base of each of the arms or lappets is a small egg shaped structure. These are the gravity receptors.

A jellyfish's normal routine in the ocean is to drift with the currents and maintain a certain depth. Jellyfish are slightly more dense than ocean water so they slowly sink toward the bottom. They maintain their relationship to the surface and light by pulsing their mantle or lappets. Even though they do not recognize left,

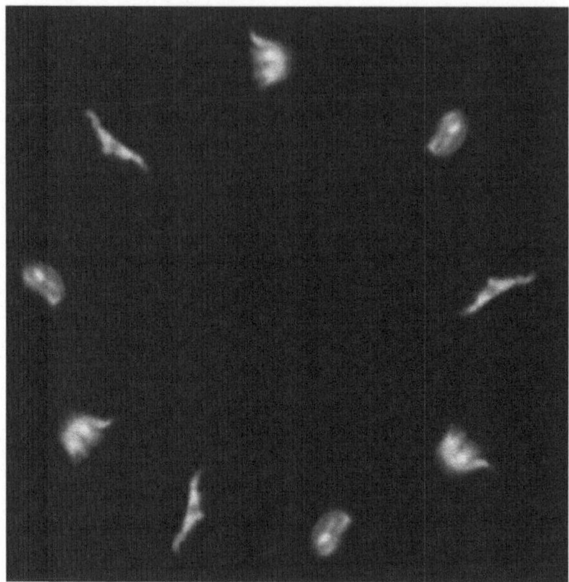

FIGURE 8.5 A jellyfish can be seen in different positions as it pulsed in a circle while in space. In the accompanying video several jellyfish swam in loops. Jellyfish never exhibit this kind of movement in Earth's gravity. See Video #004 Jellyfish. (Video courtesy of NASA)

right, backwards or forwards, they do distinguish up and down and can float upright at a preferred depth.

When in space, the young jellyfish swam in circles (Figure 8.5). Since they don't drift down or sink while in space, there was no need to pulse but they erratically did so.

Fish called Medaka, which are part of a group of fish known as killifish, have also been flown in space to study reproduction. This particular fish was chosen because of size and method of reproducing. Medaka are tiny minnow-like fish native to rice fields in the orient. The fish that flew in space came from Japan. Small fish were needed because of aquarium size restraints. As a prelude to space flight, a number of the Medaka were flown on μg parabolic airplane flights in Japan. Most of the Medaka exhibited a rapid looping, spiraling, swimming behavior while exposed to μg (Figure 8.6 and video). A few of the fish did not loop. Some of these latter fish were selected and bred and a strain of Medaka that does not loop in μg was established to be used on the actual space flight.

FIGURE 8.6 Medaka fish in μg. These fish were looping very rapidly, about one loop per second. The video pictures were taken on a Japanese aircraft during parabolic microgravity. See Video #005 Medaka looping. (Courtesy of Kenichi Ijiri)

From the non-looping strain, four fish, two males and two females, were chosen for the space mission. They successfully reproduced as we'll discover in Chap. 9.

Medaka's looping behavior wasn't the first time such behavior was seen in small fish in space. The second flight crew that visited Skylab in 1973 took a plastic bag aquarium to space divided into two compartments. One contained two small adult fish and the other fertilized eggs from the same species. The plastic bag aquarium was attached with velcro to a wall for the duration of the trip (Figure 8.7). Fish behavior was recorded twice using a video camera. The first video was recorded shortly after arrival on Skylab before they had adapted to the space environment. The fish, Mummichogs, turned loops like Medaka when they first arrived at Skylab. Eighteen days after the first video and 19 days after arrival at the laboratory, the fish had stopped swimming in loops. Instead, they

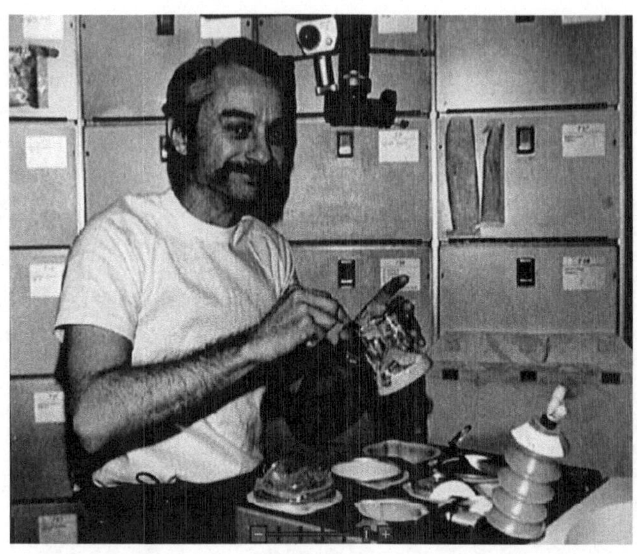

FIGURE 8.7 The second crewed expedition to Skylab was in 1973. Astronaut Owen Garriot is in the *center*. Behind him in the *left* of the picture is the plastic aquarium with the two Mummichogs, fastened to the vertical wall. See Video #006 Skylab Fish. (Photo and video courtesy of NASA & R. Simmonds)

swam with their backs to the light in the laboratory away from the compartment wall. To the fish, the wall was down and light in the laboratory was up. Just like light from the Sun indicates up on Earth. To the crew the compartment wall was vertical, as clearly shown in the picture. These tiny fish had successfully adapted by relying on visual, not vestibular, evidence after 19 days in space.

Frogs have also been chosen for space flight. Frogs change as they grow and mature. They behave in one way as tadpoles and another way as adult frogs. The frogs chosen for spaceflight were of the genus Xenopus. They are aquatic and native to sub-Saharan Africa. When they are exposed to μg on parabolic flights, adult frogs extend their limbs looking for a solid surface (Figure 8.8). There is no information as to whether adult frogs in water would loop, or extend their legs, if they were in space. Tadpoles are only swimmers and when they are exposed to μg they turn loops in a manner analogous to Medaka and Mummichog fish. In both fish and tadpoles, looping only occurs during exposure to the microgravity environment.

Swimming in water and flying in air represent life in very different environments. Yet conceptually there is a similarity

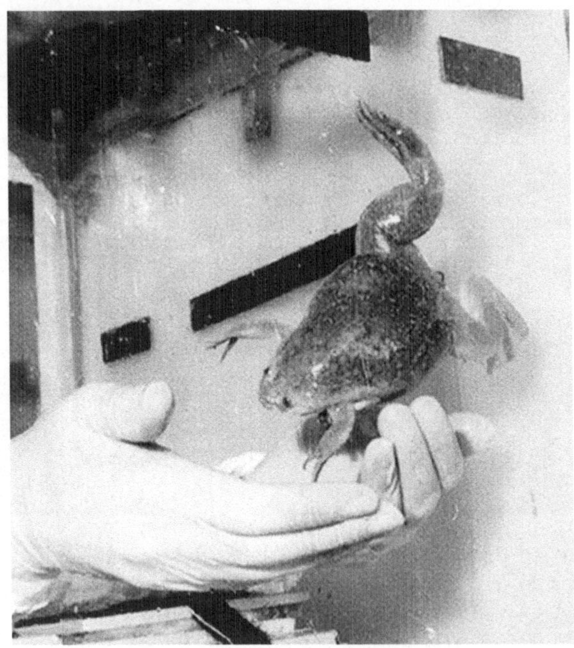

FIGURE 8.8 Adult Xenopus frog during parabolic μg. Legs are extended, searching for a solid surface. They do not loop. The accompanying video is of Xenopus tadpoles on a KC-135 flight with a number of brief periods of μg (Photo courtesy of author). See Video #007 tadpoles. (Video courtesy of NASA & K. Souza)

between flying and swimming. In both cases, there is no contact with solid surfaces.

When birds have been observed during exposure to μg, they also move in circles. However, birds have a tighter somersaulting behavior pattern than the loops that are typical of swimming animals. Pigeons were flown on the KC-135. The behavior of a pigeon released during a period of μg is shown in Figure 8.9. This sequence of frames from the video are very closely spaced in time because the bird was rotating very rapidly. The pigeon maintained its vertical orientation and did not rotate on its side. Perhaps this was due to the lights in the KC-135 which establish an up and down.

Fertilized Japanese quail eggs were carried to the Russian Mir Space Station. When the chicks were newly hatched, their behavior was very similar to the adult pigeon as they rapidly rotated in place. These unfortunate newborn birds were unable to cope with a μg environment (Figure 8.10). A video segment of an unrestrained

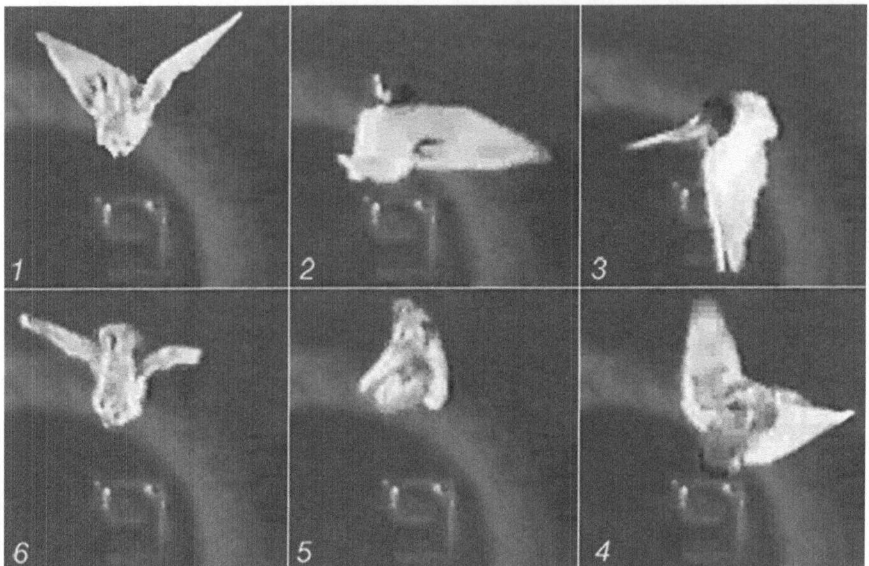

FIGURE 8.9 An adult pigeon on the KC-135 was released during μg. It flew in tight circles and rotated very rapidly. See Video #008 pigeon. (Video courtesy of NASA)

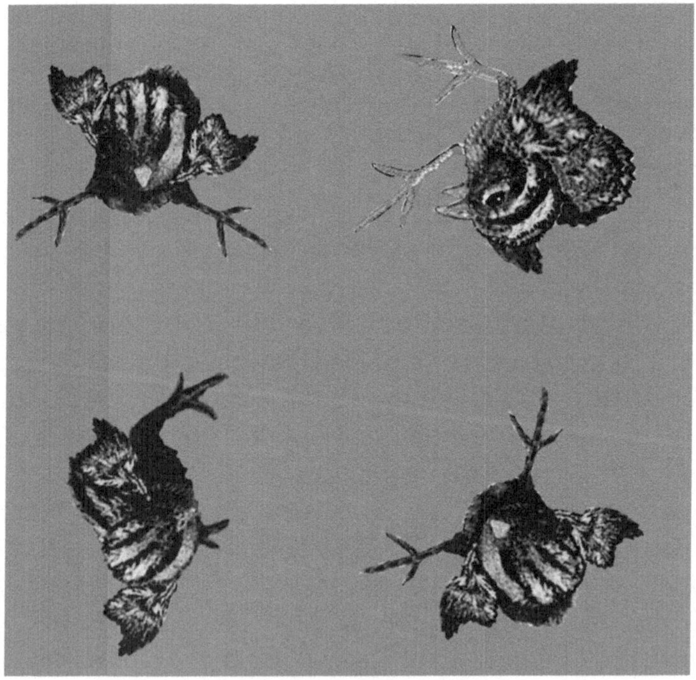

FIGURE 8.10 Japanese quail chicks that hatched on the Mir Space Station were unable to regulate their body position. Video #009 is of a single quail chick trying to orient itself. (Video courtesy of Institute of Biomedical Problems, Russia)

bird demonstrates its uncoordinated behavior. If restrained by a cosmonaut, they would peck food from his wrist and appeared quite normal.

Looping and somersaulting while in μg is a unique behavior pattern seen only in swimmers and flyers, not ground dwelling animals. Why it occurs is unknown. It may be due in part to the same type of sensory conflict seen in humans when first exposed to space flight. The Skylab fish adapted to μg and began to use visual cues to orient to light. Unfortunately, birds have not been exposed to μg for a long enough period to allow adaptation to occur. We need to determine what other changes occur in animals as they cease to turn loops and somersaults in space.

Adaptation to become better fit to live in new environments is a hallmark of life. Adaptation has been seen in all vertebrate species that have been taken into space, with the exception of birds that have only spent very short periods in μg. It will be interesting to see to what extent they are able to adapt when they spend longer periods of time in space. It may be necessary to restrain them until their neuro-vestibular/visual systems can adapt to life in a μg environment. Based on changes in other species, they will probably be successful.

Part C: The Personality and Perversity Tango

The leader can never close the gap between himself and the group. If he does, he is no longer what he must be. He must walk a tight rope between the consent he must win and the control he must exert.

Vince Lombardi (1913–1970), Football Coach of the Century (ESPN)

Absence may make the heart grow fonder but too much close contact can cause problems. Spending a number of months on a space station can sometimes make what would be minor frictions on Earth become major disagreements. You can't just go away by yourself till things cool down. There is no "away" on a small space vehicle. At the end of the work day, you already are home.

Interactions between crew members is a complex issue. It gets more involved and intense the longer the mission and the smaller the size of the crew. There is probably some optimal crew size and usable living space where potentially serious psychological problems will lessen. The factors that increase the likelihood of interpersonal conflict include: length of time in space or isolation of any kind, number of crew members involved, leadership qualities, size of the habitat or living and working space, and the behavioral/psychological makeup of the individuals on the mission. Mission length and the stability of that length are also important. A sudden and unanticipated change from 6 to 9 months away from home can result in psychological problems. This is particularly true when the decision to change mission duration is made by others. Some scientists have suggested that important considerations on international missions include tension, cohesion, leadership role, and cultural differences.

The role of the leader is one of the most critical aspects of long duration missions that are likely to be multi-national and perhaps include both genders. The effective leader must be able to change leadership style depending on the situation. The leader must be competent, experienced, optimistic, and capable of filling many roles. For example, he or she needs to clearly communicate, work to promote group harmony, yet be available and supportive in order to solve personal or individual problems.

There is an additional factor. Simply put, it is job satisfaction. As space flights become more routine, scant public attention is paid to each mission. Not that astronauts become astronauts to gain public admiration, but since the earliest human space flights, an aura has surrounded those who have traveled in space. The more common space flight becomes, the less notable it will be. Space tourism will further decrease the allure of space. There is nothing magical about being able to buy a space voyage. Astronauts, as a group, are strongly motivated to succeed. They have achieved success in their own professional careers before being selected as astronauts. As more and more tourists travel in space, that won't necessarily be the case.

The typical space shuttle flight is about 2 weeks long, which is usually not enough time for serious personality problems to develop. On these short missions, the crews have such a tight schedule there is little time available for anything but mission success. The crews are focused on completing the assigned tasks. At this time, the brief space tourist jaunts into LEO that have occurred and are being planned are much too short for negative personality conflicts to be a major factor.

There is some information on how small groups react to isolation based on Arctic and Antarctic outposts and with ships frozen in the ice over winter. Longer space flight missions like Skylab, Mir, Mir-Shuttle, and ISS have added to those early experiences. With longer periods of isolation, interpersonal problems are more common. In the space programs and also in the Arctic and Antarctica, there is a high degree of selection to have individuals who can readily adapt to others' personalities. The Soviet/Russian space program has focused on long term space missions. As part of that program, they have paid close attention to the psychosocial aspects of long term space missions. Cosmonauts are continuously monitored for signs of psychosocial changes by a psychological support group that oversees them from the beginning of training through space flight and reintegration into their community following return from space. Unfortunately, careful selection alone is not sufficient to prevent personal grievances and dissatisfaction with others' performances.

The story is told of two Russian cosmonauts on the Mir Space Station who had a falling out during their mission. One took refuge in vigorous exercise as an outlet for his feelings, spending most of his waking hours using the exercise equipment that was available. By venting his frustrations on the equipment instead of his companion, no untoward incidents were recorded. One side benefit of his activity is that he returned to Earth very fit without the usual decrements in physical status usually seen in long term missions to space. Based on this one singular example, it is potentially possible to prevent most muscular atrophy and physical adaptation. This will undoubtedly remain as an isolated example as it is unlikely that a regimen of 8 or more hours of daily exercise would be tolerated by future space voyagers as a viable countermeasure to musculoskeletal atrophy. This story reemphasizes the point that for a countermeasure to be effective it must consume a significant part of the day.

Two other factors are important: the size of the group and the size of the working and living facility. There is not a clear recommendation that the crew size should be a certain number with a specific amount of cubic living and working space needed in order to decrease the chance of interpersonal conflict. However, bigger is better. If crew size is sufficiently large, individuals can have a degree of selection of whom they desire as companions or confidants. This is a beneficial approach.

Privacy on board the new transit vehicles and new world habitats will require careful design planning and use. Individual crew members need separated sleeping quarters where they can have both physical and visual privacy. They will need to have a personal place where they can display and cherish memories of home. Televised communications with family on Earth should be private, not monitored by technicians in a computer facility or by mission supervisors. In today's world, space vehicles are noted for being small and constrained in size. Until the day is reached when the cost of launching is not as high and less more spacious habitats are available, crew size and available space will require that individuals must be accommodating, willing and able to work closely together with harmony and a relative absence of conflict.

Although disagreements have developed between crew members during space flight, the more frequent case is that there are problems between the crew in space and the ground control personnel who are responsible for managing the conduct of the flight. On more than one occasion, the crews in space have ceased radio contact with the ground for periods of time. These episodes have primarily been due to the perception by the flight crew that unreasonable requests were being made and that the ground observers did not fully appreciate the situation faced by those in orbit.

As long as space voyages are in LEO and do not leave Earth's vicinity, there can be a rapid return to Earth, if necessary, for physical or mental problems. Crews do not have the same feeling of isolation in LEO that will occur when space travel is to another planet. That will begin to change when there are trips to the moon for longer stays. Returning home to Earth will require several days of planning and then a relatively short space flight to reach Earth's vicinity. Mars presents an even more daunting task. The shortest distance would be 40–50 million miles one way and require approximately a 6 month transit time. Plus, the even more restrictive necessity, Earth and Mars must be in an appropriate relationship to each other with regard to their orbits around the Sun in order to start a return trip.

What is the difference between being involved and being committed? If you have had ham and eggs, the hen is involved and the pig is committed. In space flight, all of the ground support legions are involved, but the crew is committed. When our space program realizes the goal of human missions to Mars, there will be a new level of commitment. After all engines fire and the Mars crew vehicle reaches escape velocity, there is no turning back. They must continue on! It would be possible to do a sling shot loop around Mars and be directed back toward a rendezvous with Earth, but the whole journey would take a year to accomplish and without landing and exploration, why bother with the trip? When the crew leaves Earth, they will know that there is no way to terminate the voyage and return to Earth. They will be committed.

With the thought in mind that, a round trip to Mars will take 2–3 years. It is crucial that crews be carefully selected, not only for their technical skills but for their ability to develop strong interpersonal relationships with minimal friction. The crew will need to operate as a single unit with clearly defined goals. Given the cost of a human voyage to Mars and the interests of a number of countries in such a venture, it is very likely that the voyagers will be a mix of nationalities and cultural backgrounds. Behavior that might be tolerated or even appreciated in one culture could represent unacceptable behavior in another. It will not be sufficient for the flight crew to recognize that cultural differences exist. They need to be ingrained in various aspects of their fellow crew members' cultures. Both the Shuttle/Mir missions and now the ISS have provided the opportunity for increased intercultural learning. These missions have provided a good chance for both Russians and Americans to learn a bit about each other. It is a great improvement over the Cold War.

Based on the makeup of current space crews, a flight to Mars will undoubtedly be a mixed gender crew. Having both sexes present on a voyage of this duration will add one more dimension to the list of potential problems. Liaisons may form and/or disintegrate between couples. It could have a negative effect on the smooth operation of the mission. These are not trivial concerns. One of the American astronauts who had spent a long period on the Russian Mir Space Station said that psychological issues are one of the most crucial problems on long missions.

A Mars trip will be long and at times tedious and repetitive. On other occasions, it will require a strong background in both theory and practical problem solving with a great deal of skill and ingenuity thrown in to solve unanticipated problems that will most assuredly develop. The ability to emotionally transition from relative leisure to work overload will be an important characteristic of crews on interplanetary missions. The relationships between the commander and the crew members must be strong. The leader will need to be not just open and communicative but unbiased and forceful in his/her decisions.

Think for a moment or two. How would you like to consider, for a few years or for the rest of your life, living in a small closed facility? A new colony on the moon or Mars, will be not too large, but big enough to grow your food in an intensive agricultural unit, engage in scientific research, and explore a new world. There won't be many of you there. Just a small band of people, each with multiple skills: doctors, nurses, mechanics, builders, chemists, engineers, farmers, food processors, cooks, sanitation and recycling experts, and handymen. Then there has to be someone or some segment serving as leaders, dispute settlers, and directors. Overall, a whole hierarchy will be formed, making up an integrated unit with a single purpose. Starting and developing a community on a strange uninhabited world where you must first build inside facilities to support your every need will be a daunting task. It will not be easy even with help from Earth to get the program jump started. Some of your fellow pioneers may come and go. They will be replaced by individuals unfamiliar with the colony's life style. These newcomers will need to be both integrated and indoctrinated as they adapt to the colony's way of life. Individuals selected for initial colonization must be sociable, flexible, and forgiving in addition to possessing multiple technical skills.

Children conceived, born, and growing up in this small tight knit community will know no other way of life and integrate naturally as they mature. It is the first generation inhabitants, the explorers, and the newcomers who will need to be indoctrinated and adapt to the community's life style. They are the ones that most likely would suffer from psychological problems as they learn a new style of life, isolated and confined.

9. An Essential Activity

The natural man has only two primal passions, to get and to beget.
Sir William Osler (1849–1919), Physician,
from *Science and Immortality*

Part A: Starting the Next Generation

Reproduction is an essential activity. Without reproduction by individuals entire species would soon disappear. They will become extinct and cease to exist. Reproduction is a consuming fact of life. Perhaps it is more accurate to say that initiation of reproduction, the sexual act, is consuming for bisexual animals and the rest follows naturally. We are all mortal: birds, bees, flowers, and us. Every creature that has ever existed has a lifespan typical for its species. To keep the system in balance, death must be as common as birth. That is a problem for the Earth today. We are out of balance, humans have overloaded the planet's ecosystem. We need to slow down, move out, away, and get a fresh start.

In reproduction we pass on our special genes and characteristics to the next generation. Males, who seem constantly ready, and receptive females must mate to produce offspring. In most of the animal kingdom, the female signals the male that she is ready by producing a powerful scent, called pheromones. A single female butterfly's pheromones can attract males from miles away. Just a whiff is enough for them to travel upwind till they reach the source and join with the female in conceiving the next generation. It's a race and the first male to arrive wins by passing along his genes.

Becoming civilized has deprived us. Potent pheromones do not work in close-knit communities. Villages, towns, and metropolitan areas are not the right setting for a horde of sniffing males trying to locate the origin of an exciting and irresistible odor. We are the poorer for it.

R.W. Phillips, *Grappling with Gravity: How Will Life Adapt to Living in Space?*, Astronomers' Universe, DOI 10.1007/978-1-4419-6899-9_9,

The success and longevity of any species is characterized by its ability to adapt and to successfully spread its genes as far and wide as possible. This is true for all bacteria, plants, and animals. Successful reproduction is the hallmark of life. From another perspective, without reproduction you will never be anyone's ancestor. The urge to pass on your personal genes is strong. In order for us to colonize the moon or Mars, either reproduction must take place on these new worlds or a continuous supply of new explorers will need to be provided as earlier colonizers age or return to Earth.

If we are really interested in human reproduction and development in space, why study other organisms? The answer is straightforward. All multi-cellular organisms, plant and animal, use similar techniques. The male gamete or sperm joins with the female gamete or egg. They fuse. Each of the gametes provides half of the DNA. Together they begin the formation of a new unique individual.

We can learn a lot by studying other life forms and begin to develop a basis for initiating human reproduction when the time is right. An understanding of the pitfalls that other organisms have encountered during the entire process of fertilization, development, and maturation in space will allow us to make informed decisions on the probability of our successful reproduction away from Earth. Information on reproduction of animals more closely related to us will be of the greatest benefit, but there is much to learn from other life forms. The basic fundamentals are the same.

When we colonize our neighboring worlds with their lower gravitational fields, there will be a desire for the pioneers to add to the colony's numbers directly rather than by just importing new colonists. This will occur not just because of numbers of replacements needed, but because we are destined to sexually interact and to reproduce. It will happen.

First things first. So far there has not been a concerted effort to study animal reproduction: sexual activity, embryogenesis, birth, development, and maturation in μg. The research that has been conducted in space has provided some interesting tidbits of information on portions of the reproductive cycle in a number of different species. In most cases, each experimenter has studied only a small part of a reproductive cycle in a single species.

Based on the information available now, the odds are that animals, once adapted to the space environment or the decreased gravitational fields of the moon and Mars, will be successful in creating a next generation.

There are some problems if we just consider space microgravity, particularly in birds and mammals. Precocious birds like Japanese quail that are able to fend for themselves as soon as they hatch on Earth, never had an opportunity to adapt. They spent their brief lives constantly in non-coordinated and aimless motion.

Rats are the only mammalian species to be studied during early neonatal life. Very young rats were unable to survive. Even if adults are adapted to μg, their progeny will be born or hatched non-adapted. They will perish before adaptation can occur or they may not develop normally.

There is probably not a direct correlation between reproductive problems in μg and the success of reproduction in the reduced gravity found on the moon and Mars. There is undoubtedly some finite level of gravity that will allow reproduction to occur in species that can't effectively reproduce in μg. So far, there is no information that would help define that level. Based on the wonderful adaptability shown by animals as they cope with μg, it would seem that even the moon's slight gravity field will be sufficient for reproduction to succeed for all Earth life. At this time this is conjecture but in a few years with outposts on the moon, we can begin to learn if Earth born organisms can successfully reproduce in a slight gravitational field. Even the moon's 0.16 gravity field is enough to clearly establish up and down. The adult that was conceived, born, grown and matured on the moon may behave and look a bit different than its parents, but as long as they stay on the moon they should prosper. A return to Earth will not be so easy.

Reproduction away from Earth is important for a number of reasons. One, it will enhance knowledge of the role of gravity in this most basic life function. We need to develop a coherent and encompassing understanding of gravity's role and how necessary gravity may be for growth and maturation. To accomplish this, all aspects of reproduction in μg, the moon, and Mars needs to be studied in a variety of species. Two, when we expand to, and live on other worlds, it will be necessary that we become self-sufficient and develop the skills to grow plants and animals for food and in the process, recycle organic materials. Successful reproduction of certain species will be critical in this scheme in order to supply food and to establish a new closed ecosystem. An ecosystem that on a small scale mimics the Earth's biosphere and integrates plant and animal components. The third reason is that as we begin to live for long periods of time or permanently in space, human reproduction will become an issue. Pioneers will desire to produce their own offspring, not just for the biological imperative to pass on their genes for future generations, but also to be able to live in close proximity with their immediate family. Their offspring will add to the colony's numbers directly instead of just importing new colonists.

As a prelude to this, reproductive function in μg must be studied to determine how both the absence of a functional or lesser gravity field will effect growth and maturation of new life. Before humans attempt to reproduce, successful reproduction of other species must be proven. The first step is to determine how μg affects all aspects of the reproductive process from fertilization through development and maturation to produce a functional adult capable of initiating a new generation. We don't have enough of that information today.

Where do we stand today regarding the overall process of reproduction in space and can we make meaningful predictions where knowledge is scant? We will start with simple single cell microscopic organisms, bacteria. Bacteria are classified as prokaryotes, meaning that they evolved before (pro) organisms with a fully developed cellular nucleus called eukaryotes. The prefix eu means good. Eukaryotes include all plants and all animals. Outside of microscopic organisms, every living thing that we deal with is classified as a eukaryote. The multi-cellular eukaryotes represent the life forms of greatest interest to humans even though we eukaryotes are greatly outnumbered by prokaryote bacteria.

Bacteria grow and multiply very effectively while in μg. Actually, in most cases, they grow faster than they do on Earth. A troubling aspect of this growth is that in some cases they can develop antibiotic resistance more readily. But how antibiotic resistance is linked to μg is not clear. At this time there have not been any recognized negative effects or diseases in crew members due to changes in bacteria while in space, but it is a concern.

Relatively simple eukaryotic animals that are classified as flatworms, planaria, can reproduce sexually. Each planarium is a hermaphrodite, containing both eggs and sperm. When two planaria mate, sperm are mutually exchanged. Both "parents" develop new individuals in their bodies that are shed in capsules. A planarium can also reproduce asexually by splitting off its tail. Both segments regenerate the missing parts and form two new complete individuals. This is a less common method.

When planaria were flown on a space shuttle mission, they increased their population approximately fourfold more than the ground control animals. They were reproducing much more rapidly in the μg environment. No basis for this increase has been established, but the flight planaria were certainly more prolific.

The jellyfish Aurelia has been used in reproductive experiments during space flight. These organisms have a complex life cycle. Adult male and female jellyfish, the floaters, start the process, producing eggs and sperm that unite. As the new individuals develop, they fasten to some object in the ocean, a rock or piece of seaweed, and grow into polyps. The upper portion of the polyp then develops and releases a number of small free swimming forms called ephyrae. An ephyrae is a new immature jellyfish. A picture of an ephyrae was shown in Figure 8.4.

Aurelia polyps were carried into μg on the space shuttle. While there, they matured and produced ephyrae that appeared to be normal structurally but these immature free swimming forms had a change in their swimming behavior. In Earth's oceans, jellyfish pulse their mantles, moving up against gravity, then slowly drifting down. Refer again to the video of jellyfish in space where they are swimming in loops. The movement of the lappets on the mantle of the jellyfish cause it to move in spurts. Each time there is a contraction of the mantle, the jellyfish moves forward. Even though they looped while in space, once they had returned to Earth, the immature flight jellyfish pulsed and swam in a normal fashion. They might have been able to complete their maturation if they had remained in space for a longer period and perhaps adapted to pulse normally. At this time, we do not know what would occur with longer μg exposure.

Immature snails have also been flown in space to study development in μg. Snails have a vestibular system that allows them to distinguish up and down. They align themselves to the gravitational field as part of their normal behavior. When the immature developing snails returned to Earth at the end of their space flight, the otoliths in their vestibular systems were larger than the otoliths of snails that stayed on the ground. There was no gravitational field to detect while they were growing and beginning to mature in space, and therefore their gravity detection system kept getting larger and larger. It was a futile attempt to detect a gravitational stimulus that wasn't there.

A number of space experiments have focused on reproduction, growth, development, and aging in the common fruit fly, Drosophila. These tiny organisms are widely used in biomedical research as a model, particularly in the study of heredity and the passing of traits from one generation to the next. Flies born in space and maintained there for a large portion of their adult life appeared normal, but with some unanticipated results. Males tended to age more rapidly based on the chemical composition of their brains and then died at a younger age than their female companions. The males formed and deposited more lipofuschin in their brains. Brain lipofuschin content is a marker of aging not only in flies, but also in humans and other animals. At this time we don't know why there was a sex-related difference. Such obvious differences in males and females have not been found in other species (Figure 9.1).

Another study involved the growth and development of brine shrimp. These tiny animals can exist in suspended animation at an early stage of their development. Partially developed brine shrimp are capable of living for long periods seemingly lifeless. It is not uncommon for researchers utilizing brine shrimp as experimental animals to keep a jar of desiccated or dried and dormant brine shrimp at room temperature in their laboratory. They appear to be incapable of further activity. When placed in a seawater solution, they become active and complete their growth and maturation. Brine shrimp were carried into space and activated by being placed in artificial seawater. They not only grew, but grew much faster than ground control shrimp placed in an identical system (Figure 9.2).

Considering these four diverse biological species: bacteria, planaria, fruit flies, and brine shrimp, there is a common theme. They either multiply or grow faster and age more rapidly in space than

FIGURE 9.1 Drosophila, the common fruit fly, has been flown on several missions. Results indicate that males age faster than females based on death rate and also the accumulation of a compound called lipofuschin in the male fly's brain. This same compound is associated with aging in mammals, including humans. The brown spots in the skin of senior citizens contain lipofuschin.

comparable organisms on Earth. It is as if the very act of living in μg occurs at a faster rate than it does in Earth's gravity. Humans or even other mammals have not been in space for a sufficiently long period to see if increased rates of development and aging will also occur in us or other more closely related species during long term μg exposure. It is impossible to determine if we will age 16 months during a 12 month journey to space. Our life cycles are much too long to detect minor changes of that sort.

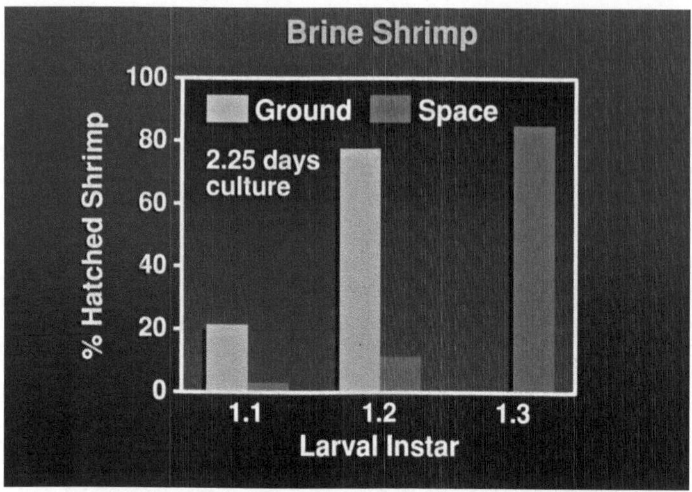

FIGURE 9.2 Brine shrimp in space grew at a much faster rate than the controls down on Earth. An instar is the period between molts. Shrimp molt or shed their outer shell as they grow larger and become more mature. The space shrimp were one molt ahead in a little over 2 days of exposure to microgravity.

What may be of greater interest to humans is the study of reproduction in vertebrates. Vertebrates are closer genetic relatives of ours and other mammals than invertebrate species like snails, flies, or shrimp. Two male and two female Medaka fish from a non-looping strain were flown in a small aquarium on the Space Shuttle. They bred on several occasions. Medaka behave differently during reproduction than most other bony fish. Fish more familiar to us have a stand-offish behavior and breed without any contact. The female makes a nest, lays eggs in the nest that may be no more than a shallow indentation in sand or gravel. The male swims over and deposits sperm, called milt, over the eggs. The male and female don't even touch or get together as we see in most vertebrates.

Medaka reproduction involves close contact. After a brief courtship, the male attaches himself to the female using tiny processes on his anal fin. He seems to embrace the female tightly, rubbing her abdomen by shaking his body. This causes the female to secrete eggs in a gelatinous mass onto the surface of her abdomen. At the same time, the male ejaculates sperm and the eggs are fertilized between the two fish (Figure 9.3).

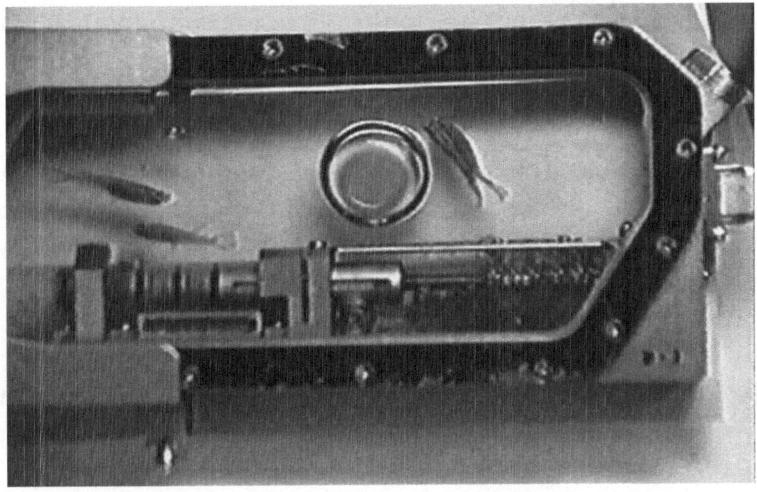

FIGURE 9.3 Two Medaka are mating while in μg on a space shuttle flight. With the stimulation of his presence and behavior, she ejects eggs in a gelatinous mass onto her abdomen. He ejaculates sperm. The eggs are fertilized and then float free to develop and hatch on their own. It is obvious that this is a space photo as the air bubble is not rising to the top of the container. See Video #010 for fish mating in space. (Picture and video courtesy of K. Ijiri)

The whole process is analogous to courtship and reproduction in birds and other animals that lay eggs. In Medaka, the fertilized eggs are released from the female, the embryos develop and hatch. During the space shuttle flight, Medaka bred on several occasions. While in space, eggs developed, hatched, and the young were returned to Earth. They matured in Earth gravity and later produced more generations of Medaka. After selecting the non-looping strain, there did not seem to be any problem associated with the reproductive cycle in these fish. Offspring of Medaka that were born in space were distributed widely to school children throughout Japan. The astrofish and their offspring were normal.

This one experiment clearly showed that at least one kind of fish, Medaka, can breed, proceed through embryonic life, hatch, and obtain nutrients as young fry. The presumption is that this same process would occur in other fish and that they could grow to become functional adults. Either non-looping strains or fish that are adapted to space for a period of time and cease looping as occurred on Skylab will be necessary. There does not seem to be any reason that fish reproduction could not occur in partial gravity as well as in μg. Fish would be a welcome source of animal protein in the diets of space colonists as well as explorers on long missions. Reproductive experiments will need to be conducted on other fish species, like Tilapia, during space flight and certainly on the moon as outposts are developed. Tilapia are increasingly popular as a firm white meat fish without a fishy flavor. They are a likely choice for space explorers to carry to new destinations.

Amphibians are closer relatives of mammals than fish. As adults, they are air breathers. Frogs of the genus Xenopus, or their fertilized eggs, have been carried into space. In one experiment, female frogs were launched into space. After reaching orbit, they were treated with a hormone that caused them to ovulate. The male frogs in this experiment were less fortunate. They were terminated on Earth preflight. The testes were removed and the sperm extracted. The sperm were sent into space on the same launch as the female frogs and mixed with the eggs following ovulation by the female. From a male's perspective, there was a great advantage to being a female in this experiment! The fertilized eggs hatched while in space, after a normal embryogenesis. There were some minor anatomical differences during early embryonic development, but the newly hatched tadpoles looked normal. As noted in Chap. 8, Part B: Looping and Somersaulting, tadpoles loop in the same fashion as other swimmers when exposed to μg.

The tadpoles were returned to Earth 2 days after hatching. Their subsequent development and maturation while in Earth's gravity field were normal. However, when the space-born tadpoles first returned to Earth they initially sank to the bottom of the water tank they were placed in Figure 9.4. Soon after arrival on Earth they would reach the surface and thereafter behaved normally like the ground-based tadpoles. Curious behavior but a simple explanation. Young tadpoles hatched in a shallow pond or pool of water on Earth swim to the surface early in life and breathe air, which inflates their lungs. This decreases their density and they float more easily. In μg with no obvious up or down, the tadpoles did not or could not swim to the surface and their lungs did not inflate until after they were returned to Earth.

In another experiment eggs of Xenopus frogs were sent into space after fertilization and the young hatchlings remained in space

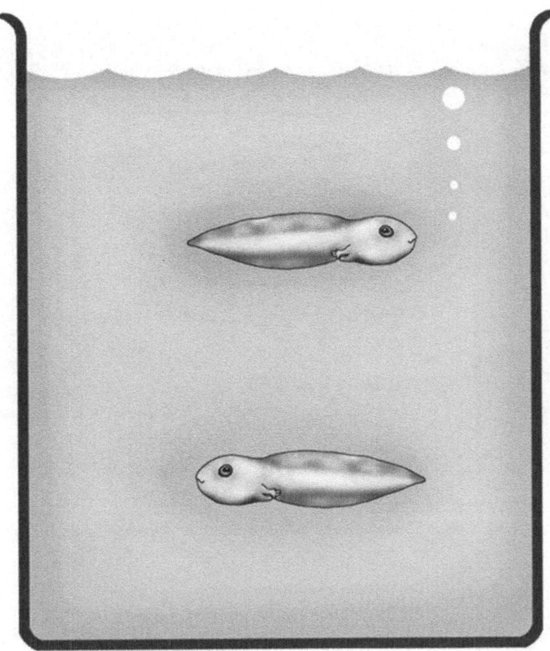

FIGURE 9.4 Tadpoles that developed and hatched in space, floated near the bottom in a column of water when they first returned to Earth.

much longer. A number of malformations were obvious when the tadpoles returned (Figure 9.5). The condition is an overextension of the back called lordosis. The tadpoles most severely affected were almost horseshoe shaped while others were less distorted. Upon examination with a microscope, other abnormalities were seen in the tadpoles that launched during embryogenesis and remained in space for a longer period. The increased gravitational load during launch could have been a contributing factor. It appears that frogs may be able to produce young in space but abnormal development could limit the new generation's chance of becoming normal, mature adults. We have yet to examine amphibian reproduction in space, or in other gravitational fields, for the entire period from fertilization to metamorphosis when tadpoles become frogs.

Birds are endotherms, or warm blooded animals like humans. Chickens and their eggs are an important part of the diet in many cultures. We need to know if chickens can successfully reproduce in space and eventually in reduced gravity fields. Several

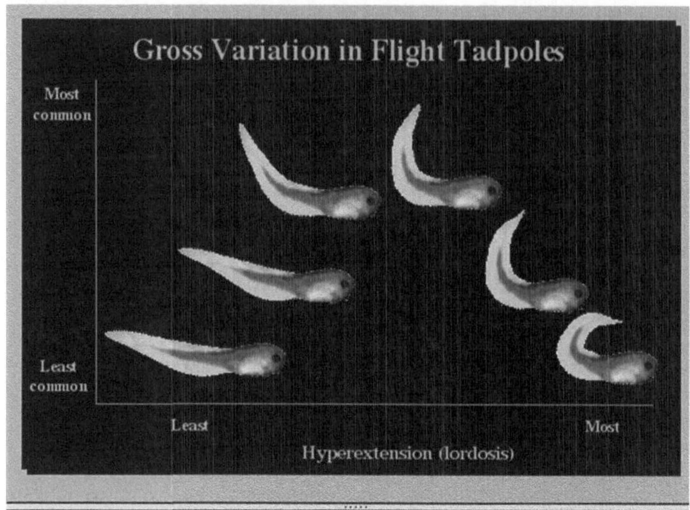

E. Snetkova et al.: Xenopus Development in Space. Journal of Experimental Zoology 273:21 (1995)

FIGURE 9.5 Tadpoles developed in space after fertilization on Earth. Many had serious developmental deformities. Most prominent was an overextension of the vertebral column called lordsis. The experiment has not been repeated. It is not possible to say whether the growth anomaly was due to μg or to launch stress during early development. Other abnormal changes in internal structure were also seen.

experiments have been conducted to determine if fertilized eggs can develop during space flights. The final results are not promising for hatchlings, although there could be ways around the problems that have been seen.

In the first experiment, chicken eggs that had been fertilized prior to flight were sent into orbit on the space shuttle. The embryos that were 0 or 2 days of age died several days after reaching orbit. The increased gravitational force associated with launch didn't seem to be a problem. The investigators think that diffusion of oxygen through the shell was too limited in μg to support embryonic development. There was no air movement in the incubator around the eggs. Once oxygen near the eggs was depleted, there would be less oxygen available to diffuse into the eggs. No mechanical mixing of air with a fan was provided, which might have lessened the problem. Older embryos, 7–9 days of age, all lived. The developing embryos were returned to Earth before hatching. Once they were hatched they were essentially normal.

A more advanced incubator has been flown on the International Space Station. The new incubator system has two compartments: one rotates to create a gravitational field, the other provides a μg environment for the developing embryos. In this experiment Japanese quail eggs were used. They are smaller than chicken eggs thus reducing the potential problems with oxygen diffusion. There was no difference in the embryos centrifuged at 1 *g* in space and those that developed on the ground. The flight embryos kept in μg were about 30% decreased in weight. This change in weight indicates that there was an inhibition of embryonic development. As planned in the earlier experiments, the developing eggs were returned to Earth before hatching. They hatched on schedule. However, the μg chicks had not grown as much as the controls. These results are completely different from the planaria, fruit flies, and brine shrimp that grew and aged faster in space. Very puzzling, but we don't have all of the answers! We are just beginning to understand how space flight changes the development of Earth life.

Embryonic development in eggs although crucial is only one aspect of reproduction. Hatching, as well as care and feeding of the young, is likely to be a more critical problem. The Russians flew fertilized Japanese quail eggs on the Mir Space Station. While in space, the quail eggs completed their development and seemingly normal quail chicks were hatched. Unfortunately, the next stage of their lives was short and unpleasant. Newborn quail and baby chicks are precocious; that is, relatively mature when they hatch. They can walk immediately after birth and begin to select and consume food to sustain themselves. They do not have to spend time being fed in a nest. Their normal behavior is to move about on a flat piece of Earth. Quail and also chickens are much more advanced at the time of hatching than many other baby birds that must be cared for in the nest by their parents.

That relative maturity did not help the newborn quail chicks' adaptation and survival when they were hatched in space. First of all, they were floating instead of walking. They flapped their short wings and struggled making flailing movements with their legs. They began to spin in a spiraling somersault fashion making as much as one revolution per second. A picture of a single quail chick on Mir was shown in Figure 8.10 and Video #009. Because of this rapid movement, they were unable to obtain food or water without being restrained. If held, they would eat food from the cosmonaut's hand, but when released and on their own, they immediately started gyrating again, bumping into each other and into the walls of their chamber.

They were aptly described by the crew as being in a state of confusion. The crew spent a considerable amount of time observing the birds, trying and failing to adapt to this unusual environment into which they had hatched. The behavior of the group of the quail that hatched in the incubator on the Russian Mir Space Station is shown in Video #011 (Video courtesy of the Institute of Biomedical Problems, Russia).

Their bodies were ultimately returned to Earth. Upon examination, they appeared normal. A quote from one of the Russian space scientists is very descriptive of the problems encountered.

> This stage of the experiment turned out to be short and for the different hatchlings 2–4 days depending on the time of hatching. During this time the hatchlings were not able to adapt to weightlessness and independently acquire the ability to attach their claws to the floor, which created problems for them to feed independently, although confidently pecking food from hands. During this time the hatchlings were continuously in a state of confusion rotating and when touching the net would not try to attach themselves. Sharp movements of wings and legs were observed which only increased the rate of rotation, which often caused greater than one rotation per second.

At least for this one species of birds, Japanese quail, completing a life cycle in space is not possible with current technologies. Some scientists have developed restraint devices to hold the bodies of newborn chicks in a fixed position. That would allow their vestibular systems the opportunity to adapt to μg before they are turned loose. At this time the restraints have only been used in Earth-based studies.

It will require a considerable effort to get birds to eat and drink while they are adapting to μg. Even if that is successful, the adaptation would not be transmitted to their offspring so that they could grow and mature. Successful reproduction of birds in μg seems questionable at this time. Reproduction of chickens or quail on the moon and Mars should be a different story. Based on the behavior of the precocious newly hatched quail chicks and adult pigeons that somersaulted on the KC –135 μg airplane, we are not ready to have birds in μg. To successfully care for birds in space it may be necessary to provide a gravity field by centrifugation. On the other hand, birds have only had very brief μg exposures. If restrained for a number of days or even weeks, they may adapt and overcome their disoriented behavior.

A solution for future colonies would be to take fertilized chicken eggs to the moon and allow the chicks to hatch there in the slight gravitational field. However, this would not solve the problem of getting birds to Mars, the 6 month trip is much too long. For a Mars trip it would be better to use centrifugation to create a gravitational field in an avian facility inside the transit vehicle than to try to force the birds to successfully adapt to μg. The final solution to the problem of carrying birds to Mars will depend upon their ability to adapt with longer exposure to μg. If they can't adapt, we will need a centrifuge facility or rotating tethered space vehicles to provide some level of gravity during the long 6 month trip.

Other mammals are closer relatives to us than vertebrates like fish, frogs, or birds. The most logical approach to follow in conducting preliminary studies on the effect of μg on mammalian reproduction is to use an animal model like rats. Rats have a short gestation period, multiple offspring, and reach sexual maturity in a matter of months. They are widely accepted as a valuable source of information about the way that we may respond. There have been a number of reproductive and developmental experiments conducted in space using laboratory rats.

Several problems have been identified while studying rat reproduction in space. They are serious enough that it does not seem likely at this time, that it would be morally justifiable to permit women to give birth in μg. The gravitational fields on moon and Mars are a different story. A few of the potential roadblocks to μg mammalian reproduction are known and the chances of success do not look promising for any of us.

One Russian experiment considered the first phase of reproduction when two male and five female rats in separate compartments were launched in a biosatellite without a human crew. Two days after launch a panel between the male and female compartments opened and the seven animals were together for the rest of the flight. Seven days later the satellite and its animals were returned to Earth. No females were pregnant. According to one scientist, there was evidence that several of the females had been pregnant and that the pregnancy had terminated for unknown reasons. Other scientists felt that no pregnancies had occurred. It remains an open question today.

Regardless of whether breeding and fertilization occurred or not, the important information is that there were no viable pregnancies. It would be a very unusual occurrence on Earth for two male rats to be sequestered with five females for a week without a single pregnancy. There are several possible explanations. One, the rats were not able to physically get together while floating. A breeding chamber that would facilitate coupling would need to be included in the rat cages. The second possibility is that due to the stress of launch and the absence of a functional gravity field, the female rats did not go through estrus and thus were not ready to accept the male. It was recently recorded that female mice in space did not go through estrus periods and would not be receptive. That makes it likely that the problem with the Russian female rats is that they did not go through an estrus cycle on their brief trip. The results from several studies have indicated that there is a decrease in rat sperm function following space flight. In other experiments, no change in sperm viability has been found. Certainly, children have been born to the wives of male astronauts following space flights of varying lengths, so that sperm changes due to μg probably are either short term or peculiar to rats.

Assuming that impregnation is successful, the next period of concern is gestation, when the fertilized egg becomes an embryo and then a fetus. Female rats, at varying times after impregnation and prior to birth, have been flown in space. Two space flights took pregnant females within 1 or 2 days of giving birth, before they were returned to Earth. They produced normal rat pups. The females experienced an increased number of labor contractions during the birthing process. This was presumably due to loss of abdominal muscle strength during the space flight. Another change was in the newly born rat pups. Immediately after birth, they had trouble righting themselves by turning over. Righting performance was normal several days after birth. Spending the latter part of pregnancy in space did not seem to be seriously detrimental to pregnant rats or to their offspring.

We come next to the most critical period in the reproductive process of mammals. This was a troublesome period in frogs and the lethal end of reproduction in Japanese quail. It is the neonatal period, when many newborn animals are immature, fragile, and unable to care for themselves. Newborn mammals are completely dependent on their mothers for nutrition. The early milk also contains antibodies to help the young survive entry into a hostile world, filled with potentially dangerous and even lethal viruses and bacteria. The newborn need their mother's milk until they can develop a competent immune system of their own.

Some aspects of rat neonatal life in space have been studied. The results have been mixed. Before venturing to space, consider how rats on Earth handle the end of pregnancy and the early life of neonatal pups as they begin to adapt to the world outside the womb. First, the female prepares a nest as she nears the end of gestation. This is a private place where the young can be born and stay in close contact with their littermates. They share and conserve body heat by huddling together. Under normal conditions on Earth, they lay in the prone position, which is on their stomachs. When it is time for the mother to nurse her young, she enters the nest, crawls over her babies with her legs spread widely. The newborn pups sense her presence and rollover on to their backs into the supine position. They can then attach their mouths to one of the nipples to gain nourishment. All of this activity takes place in the nest, particularly in the early postnatal period. Rat pups are immature when born but mature rapidly. About day 13, their eyes and ears open. Before that, all of their interactions with the environment and their mother are by touch without vision and hearing. They are usually weaned by day 21 and sexually mature by day 50, ready to start a new generation.

If there were no deaths and all females started the next generation at age 50 days when they became mature, the possibilities are astounding, or perhaps frightening. Two rats, one male and one female, could develop a colony of thousands and thousands of rats in a single year. This is based on an average litter size of 10, which is fairly normal, with half males and half females. Each new female would join the reproductive assembly line by birthing her

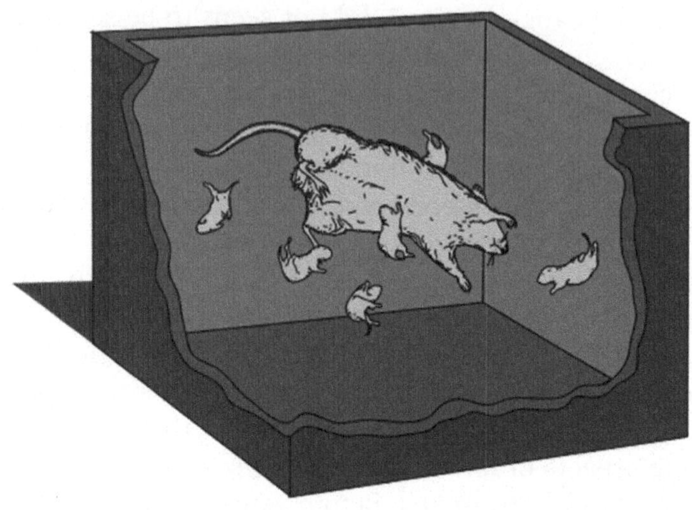

FIGURE 9.6 A nest, suitable for newborn rat pups in space, has yet to be designed and flown. The accompanying video (Video #012) shows the behavior of rat pups of several ages while in space. (Video courtesy of NASA and NIH from STS 72)

first litter at 70 days of age. No wonder rats are so common and numerous all over the world.

What happens if we try to play out a birthing and neonatal scenario in μg? Frankly, it is a disaster for the very young and immature. First, consider the nest. The rats, mothers and pups, were acclimated to a nesting area prior to launch and it appeared to be satisfactory. In space, the young pups were no longer in the nest or in contact with their littermates. Instead they floated randomly around the cage bumping into the walls, each other and their dam (Figure 9.6 and Video #012). The mother would attempt to retrieve them, but was rarely successful. Not being able to huddle and gain heat from their siblings was worsened by the inability of the mother to effectively nurse a scattered grouping of floating and shifting young. They were rarely in stable position as their orientation was continuously changing. Caring for the very young was an impossible task. The experiments conducted so far indicate that young rat pups cannot adapt to the μg of space.

No experiments so far have included birthing and the first 5 days of life. With the best habitat available at this time, few animals that are 5 days old at launch survive. Based on what we know now, birthing and living through the first 5 days will not be successful for baby rats. At 8 days, there is a 90% survival rate and 100%for the pups that were 15 days old at launch. This should be put into some degree of perspective in comparison to human infants. Rat pups learn to walk at day 14 or 15 after birth and are considered able to fend for themselves when weaned about 3 weeks after birth. No rats have given birth in space while living in a centrifuge. It seems likely that such an approach could be successful if the mother and pups stayed in the centrifuges's gravitational field until they were weaned.

In contrast to rat pups, human infants learn to walk at about 1 year of age. Before that, they are completely dependent on their mother. They cannot effectively fend for themselves without maternal input until they are probably 6–7 years of age, or more. At birth, we are much less mature from both a physical and a mental state than other vertebrates that have flown in space. It is difficult to imagine the many problems that mother and child would face if an infant were born into a μg environment. Even if feeding and other care could be accomplished, there are other challenges to be considered.

For example, rat pups born on Earth have been sent into space during their early developmental period. If they were in μg 14–15 days after birth, problems were apparent when they returned to Earth. That age is the time when they should have learned to stop crawling and to walk. They could not gain that experience because of the μg environment as they were floating in free fall. After return to Earth, they did indeed learn to walk, but their movements were not normal. They took exaggerated steps, lifting each leg higher than normal. Much later after adapting in many respects to life on Earth, they still did not walk normally. This is one example of the fact that there are windows of opportunity when newborn animals acquire skills that will be used throughout their lives. If a child were born in a μg environment and lived there for several years, during the periods when they should normally acquire gravity-related physical skills, they would likely have a great deal of trouble fully adapting to life on Earth or other gravitational fields as they mature.

Based on the reproductive problems identified at various stages in other vertebrates, it is not just prudent but mandatory that a gravity field will be required for many portions of the human reproductive cycle. Will a full Earth gravity field be necessary for reproduction to succeed? Will a child developed in a lower gravitational field ever be able to move to the parents' home planet? These are not fanciful questions. Current plans are to return to the moon with its one sixth gravity field and establish an outpost that may well turn into a permanent colony. When such a colony is inhabited by long-term residents, partnerships and marriages will take place. The desire to pass on genetic individuality is imbedded in living systems. Pregnancies will be initiated by the colonists. Will it be necessary that the mother-to-be return to Earth to complete her pregnancy? Will she need to remain on Earth until after the child is born and has developed necessary gravitational skills? At what age should a developing child be allowed to move to the moon from Earth to become a lunar resident? These are all unknowns at this time. Information must be acquired on the reproduction of other mammalian species in a lunar environment before we take these steps. A different question also needs to be asked. If the parents plan to have a permanent life on the moon and desire that their offspring should grow and develop there as well, who is to tell them no? Such individuals may never desire to return to their parents' planet of origin.

On a longer term, many are convinced that we will be successful in reaching Mars and that colonies will be established there. Again the question needs to be asked, will the Martian gravity field be adequate to develop humans that can easily travel to and adapt to planet Earth's gravity?

In summary, although human reproduction in μg does not currently seem morally justifiable, more knowledge must be gained before decisions can be made about other worlds and unfamiliar gravitational fields. At this time, there is no reason for supposing that Earth species cannot adapt, prosper, and reproduce in a lesser gravity. This environment could have the gravity of the moon or Mars.

In both of these locations, the question to be asked is whether new generations can successfully return to Earth. Earth is the birthplace of our race and the desire to be able to return "home" will be strong.

Part B: Seed to Seed

From the sexual, or amatorial, generation of plants new varieties, or improvements, are frequently obtained; as many of the young plants from seeds are dissimilar to the parent, and some of them superior to the parent in the qualities we wish to possess... Sexual reproduction is the chef d'oeuvre, the master-piece of nature.

Erasmus Darwin (1731–1802), Grandfather of Charles Darwin, Botanist, Naturalist, Inventor, and Philosopher (*Phytolagia, 1800*)

To most people, sex and reproduction relates to animals, not plants. That is a narrow view of reality. Many plants are hermaphrodites and contain both male and female components. They may self fertilize or cross fertilize. The formation of an egg or ovum and a sperm or grain of pollen is required in the reproduction in both plants and animals. They must join together to initiate the formation of a new individual, starting the next generation.

Reproduction, growth, and development of plants in space or on other worlds is important, not just to understand reproduction in μg or reduced gravitational fields, but to supply food for space travelers. On really long voyages, or for living on other worlds, importing sufficient food for the crew is not an option. For supplying food to the ISS or an initial outpost on the moon and even for a 2–3 year round trip exploratory expedition to Mars, we should be able to transport sufficient stable food to support the crew. Longer trips or a permanent residence on the moon or on the Martian surface requires a different approach. We must learn to grow foods in intensive closed systems. Closed systems for both living quarters and food production will define our new life style when we leave Earth for the moon or Mars.

The habitation modules on other worlds will be pressurized with temperature and atmosphere control. They will be small at first, perhaps analogous to a submarine with an attached greenhouse, but expanded to meet the needs of a growing population. It will be a new and different life, but we can do it. A benefit of growing crops for food consumption on other worlds is how easy it is to transport seeds compared to transporting food. No special requirement or technology is needed to send a plentiful supply of seeds ready to be planted on other worlds. Then they must grow, mature, and be harvested. It will be necessary to process, cook, serve, and consume the final product. If you are going to live on another world, don't ever forget the last absolutely necessary step. You must save and recycle all organic wastes without exception. This requirement will be especially important on the moon where there is no natural source of nitrogen and only limited carbon. These elements must be imported and recycled.

Seeds are remarkably stable and can remain viable for long periods of time. Seed storage in space started in April 1984, when a NASA space shuttle placed the Long Duration Exposure Facility (LDEF) in Low Earth Orbit. It was planned that it would be returned to Earth by a space shuttle flight in 11 months. It contained, among other things, tomato seeds provided by the Park Seed Company. Due to delays and then the Challenger accident, LDEF was not returned to Earth until January 1990. The seeds were exposed to the space environment, including radiation, for 5 years and 9 months. and the seed produced normal tomato plants for the investigators as well as students at many schools throughout the country who received samples of "space tomato seeds." Sending seeds in cargo ships to either the moon or Mars should not prove to be a problem. LDEF also carried bacterial spores. They were also viable after their long exposure to space conditions.

In many of today's affluent first world societies, people are spoiled by the overabundance of foods choices. It will be a different matter when we begin to grow food under intensive agricultural systems with limited space and a small number of selected plant species. The ingenuity of food preparers, particularly those involved in creating new recipes, will be challenged. First things first, can plants grow and reproduce in space or in a decreased gravitational field? What plants will be important? At first, it will be those plants that are necessary for our nutrition and the nutrition of the animals that accompany us to new worlds. A few esthetically pleasing plants will also be an important component. Combining the esthetics and nutrition will be a double benefit. Seeing strawberries or tomatoes grow and ripen will be both a delicious treat, when harvested, and a familiar remembrance of watching them grow on Earth.

There is a great deal of similarity between reproduction in plants and animals . Each group has evolved very effective ways to join the two gametes, egg and sperm, to start the next generation. The process of fertilization, called pollination in plants, requires both a female and a male component. They must both be ready to interact at the same time. Egg and sperm are represented in plants by the ovule or female germ cell usually at the center of the base of the flower, while the pollen grains represent the sperm. The male and female components come together in the ovary. This sequence, whether by self pollination, the wind, an insect, or other animal pollinator, is the method by which the hereditary information (DNA) in the pollen grain is carried to the ovary and its ovules. This results in the initiation of a new individual plant. The fertilized ovum undergoes embryogenesis just as it does in animals after fertilization. In plants, it is called seed formation. At this point, there is a significant difference between plant and animal reproduction. In plants, the reproductive process is complete with a viable seed that will restart its growth when environmental conditions permit. A mature seed contains a new individual plant in embryonic form. It is the next generation of its species, capable of remaining dormant for long periods. In animals, reproduction is continuous, an uninterrupted development from fertilization to adulthood.

Plants in nature have evolved many methods for ensuring that their seeds will be transported to a new location where they complete their life cycle and spread their species to these areas. Some have light and airy seeds that can be carried easily by a breeze to a new territory. Think of dandelions, a supremely successful spreader. Oak trees drop acorns straight down. Squirrels pick them up, move them around, "plant" them for their own future use and then forget the location. Squirrels are an important component of the growth and spreading of oak forests. Some seeds have burrs and stick to animals' fur or people's clothes to travel to new sites. Others are eaten, pass through animals' digestive systems intact, to be later dropped in new locations. Successful plant species have evolved in lots of ways to spread their genes. None of these techniques will be needed or used in intensive plant growth facilities in space or on other worlds. Plant ingenuity in seed distribution won't be needed anymore. Pollenization away from Earth could be a problem for those species that have evolved to depend on animal pollinators. Perhaps bee colonies will need to be part of future plant growth habitats on the moon or Mars. Bees are important animal pollinators on Earth. They produce a delightful product, honey, in addition to pollinating.

Plants are gravitrophic, that is they respond to gravity. They are also phototrophic and respond to light. Shoots grow upward against the gravity vector that is directed towards the center of the Earth. Upward growth can be modified by the presence of sunlight or artificial light, as plants are also attracted towards a light source. Roots grow down in the direction of the gravity gradient to obtain nutrients and water from the soil. Right there at the surface of the ground, plants have a 180° change in behavior. Their shoots grow up and their roots grow down, an amazing but very effective system. That marked change in orientation has existed as long as plants have been present on Earth. In μg without a functional gravitational field, plants will have lost the major signal that affects their growth on Earth. They will need to adapt to a strange new world.

The downward growth of roots is believed to be due to the presence of small particles called statocysts in the root tip cells. These statocysts produce a directional response by resting on the bottom of the cell. They are heavier than other cell components and settle here in Earth's gravitational field. In μg, weight is irrelevant. The statocysts no longer rest on the bottom directing root growth (Figure 9.7). This results in a random root growth in μg. The direction of root growth is related to nutrient and water availability, not gravity. Shoots continue to exhibit phototropism during space flight. Roots without a gravity vector grow in a more erratic fashion. Early in the space flight era, it was determined that seeds could germinate and grow in space, but the missions were too short to permit maturation and seed development.

Most American space missions have been 1–2 weeks or even several months in length. This is not enough time to allow our well characterized crop plants that "feed the world" to progress from seed to seed. That is, planting a seed and harvesting the next generation of seeds is necessary to complete a reproductive cycle. For the most part, the study of plant reproduction in space has been carried out using laboratory plants with a very short development period. More recently some experiments have investigated the life cycle of wheat. Wheat, particularly miniature wheat, is a likely candidate to be used to supply food for astronauts.

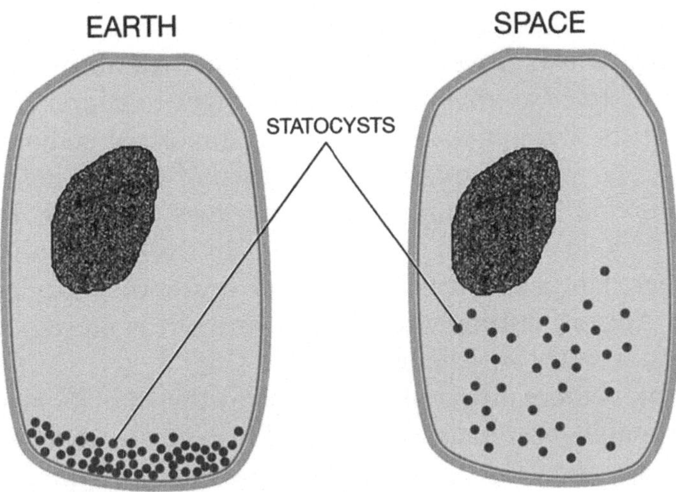

FIGURE 9.7 Statocysts in root tip cells are believed to direct root growth by putting pressure on the lower cell membrane.

On the Russian Mir Space Station and the International Space Station, a series of plant reproductive studies have been completed using primarily short generation-time laboratory plants. Each new experiment was built on the results of a previous one. The overall experimental plan was to determine if plants could germinate, grow, and mature through two generations from seed to seed to seed.

Plants face different problems in growing and maturing than what was found with animals. The first plant chambers used in space for reproductive studies were enclosed facilities that provided a normal atmosphere as well as water and nutrients in a growth media. The plants grew but did not produce viable pollen grains or ovules. Thus, they could not develop seeds. Based on the rate of growth and degree of maturity, this problem appeared to be due to a shortage of carbon dioxide (CO_2).

On Earth, outside or in green houses, plants are dependent on air movement for circulation of oxygen (O_2) and CO_2. Without air movement, plants cannot breathe properly. Carbon dioxide is the waste gas of respiration for nearly all life, although we tend to think of it as peculiar to animals. Plants utilize this carbon dioxide and water when exposed to artificial light or sunshine to produce carbohydrates during photosynthesis. That's the origin of the word carbohydrate. It is the result of the process that plants use to combine carbon dioxide (carbo) and water (hydrate) to make this important part of animal diets. Without light to trigger photosynthesis, plants will not produce these all important carbohydrates. Instead, they will use oxygen as a source of energy and produce carbon dioxide just as animals do. This goes on in the dark portion of every day and to a greater extent in the northern latitudes (December, January and February) and southern latitudes (June, July and August) during winter when sunlight is present for only brief periods in the middle of the day.

Air movement on our planet occurs by the wind for plants outdoors. If inside, thermal gradients and fans create air movement. Hot air rises and cold air sinks as is easily seen by the efficiency with which hot air balloons function. On Earth, in closed plant growth chambers, thermally induced air movement or convection is important to ensure that "fresh air" is always coming into contact with the surface of leaves to support their growth and development. In space, there are no thermal gradients since temperature differences do not cause air movement. Therefore, there is no air mixing. Refer back to the burning candle in Figure 5.5 for a clearly illustrated image of the lack of movement of heated air while in μg. Under μg conditions, the CO_2 near the leaf surfaces is utilized by the plant, but without air movement, sufficient additional CO_2 is not available and growth becomes limited. In addition, within the flower, oxygen is required for the development of reproductive structures. Oxygen deficit occurs with the lack of convection. Plants grown in closed chambers, while in μg, had shrunken and malformed pollen grains and ovules. No seeds were produced.

In a second experiment, the level of carbon dioxide in the plant chamber was increased. The plants grew better, but still did not produce viable seeds. A third experiment pumped air from the crew cabin through the plant growth chamber. The pump ensured that air was continually being mixed. This movement provided CO_2 and O_2 at the surface of the leaves and the flowers. There was an additional benefit to this approach, particularly if conducted on a larger scale. Because of the energy cost of removing the crew's exhaled CO_2 from the space vehicle's atmosphere, this gas is kept at a concentration much higher than found on Earth. Here we breathe air that is 0.03–0.04% CO_2. On space vehicles, it is often in the range of 0.4–0.5%, a greater than tenfold increase. Having plants utilize some of the excess human CO_2, not only benefits us and the plants, but is the beginning of a functional biological life support system.

On Mir, the plants in the higher CO_2 atmosphere that was continually being mixed, grew and flowered producing viable seeds that were, in turn, planted to produce a second generation of plants. These second generation plants were smaller than the parents but indicate that it is possible for plants to successfully reproduce in space. Remember these were fast growing and rapidly maturing laboratory plants, not the kind that will be candidates for producing human food in the future (Figure 9.8). More

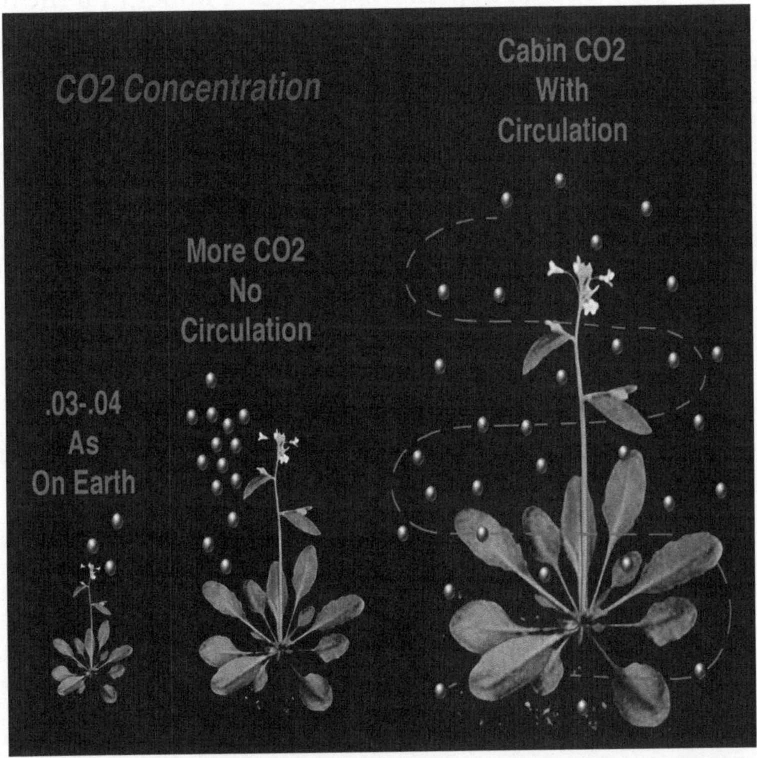

FIGURE 9.8 To study plant growth and reproduction in space, three different experiments were done at three different times. (**a**) Low levels of CO_2 as found on Earth. (**b**) Additional CO_2 was added to the plant chamber. (**c**) Even higher CO_2 from the crew cabin, with mixing, ensured that CO_2 came in contact with the leaves as it would on Earth. This stimulated plant growth and allowed production of viable seeds for the next generation. (The artistic credit is to Dennis Giddings)

recently, several long duration space flight experiments have demonstrated that wheat can grow, mature, and produce seeds during space flight.

It seems reasonable to believe that remaining problems associated with plant reproduction in μg can be solved. In the future, we will be able to grow plants in μg and provide fresh salads and vegetables to space crews. The limiting factor is the decision to devote precious space facilities and crew time for this purpose.

There are still many questions to be answered regarding plant growth and reproduction in lesser gravity fields. Not even a first experiment has been conducted. Based on the fact that plant growth and reproduction can occur in μg, there should be no serious problems with plant reproductive biology on the moon or Mars. Importing carbon and nitrogen to the moon to support plant growth will be a logistics problem, but that is a different story.

Part III
Preparing for the Future

10. Living Away from Earth

The Newtonian principle of gravitation is now more firmly established, on the basis of reason, than it would be were the government to step in, and to make it an article of necessary faith. Reason and experiment have been indulged and error has fled before them.

Thomas Jefferson (1743–1828), Scientist, Philosopher, Statesman, Architect, 3rd President of the United States, Author of the Declaration of Independence, from *Notes on the State of Virginia 1781–1785.*

Part A: The Long Haul

Reproduction and the development of a 'normal' next generation is one important hurdle for moving away from Earth; however, it is just the beginning. Nothing is known about long term multi-generational affects, of not only space µg, but in the reduced gravity fields of the moon and Mars or other planets. Another unknown is the long-term effects of extraterrestrial radiation; solar radiation from the Sun and cosmic radiation from outside our Solar System. Until we begin to live on other worlds, we must rely on conjecture and best estimate from what is known. Although radiation is a big biomedical concern, it can be blocked by insulation or deflection while in space. Alternatively, it can be avoided by locating living/working quarters under the surface of the moon or Mars to provide shielding. Living underground, settlers will be safe from radiation. This approach can work for living and research quarters, but explorers and settlers will still need to venture out on the surface. There they will be exposed to increased levels of radiation.

Since we don't yet have experience living away from Earth, we need to prepare for adaptations that will occur as we respond to strange new environments. We have much to learn. Is there any way to make logical predictions regarding changes due to life-long or multi-generational exposure to µg or to the effects of other

R.W. Phillips, *Grappling with Gravity: How Will Life Adapt to Living in Space?*, Astronomers' Universe, DOI 10.1007/978-1-4419-6899-9_10,

gravity fields? Prediction of anticipated behavior or adaptations is possible in many cases. The basis for this assumption is that a continuum of response of biological systems to different gravitational fields has been established. This may be linear as depicted in Figure 10.1. If we have information on the changes of a particular body function in μg, 1*g*, and hypergravity as represented by the centrifuges generating a 2*g* or 3*g* field and if those responses are reasonably linear, we can make predictions of what the response will be in intermediate gravitational fields. Some of that information is available now. For ourselves and many other species, informed estimates can be made about adaptations that will likely occur during exposure to other gravitational fields. The ability to do this will be particularly important as we send explorers back to the moon for months, years, or generations.

As information is acquired on the moon, it will facilitate future predictions as we send exploration parties, then colonization missions to Mars or to moons of other planets in our Solar System. We don't have a lot of information concerning generational

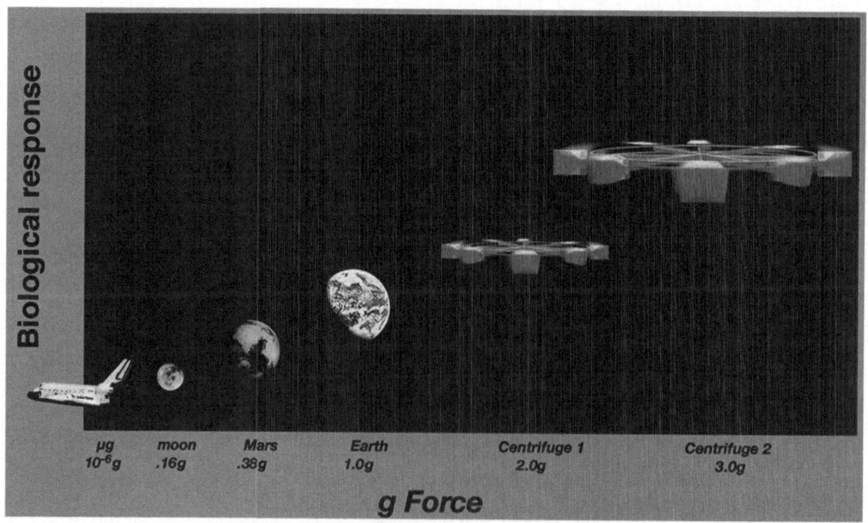

FIGURE 10.1 Considerable information from experiments on fellow humans as well as other animals supports the notion that there is a gravitational continuum. If we have data regarding adaptations in μg and hypergravity, as well as Earth gravity, and those responses are linear, we can make reasonable predictions about changes that will be seen in intermediate gravity fields on the moon and Mars.

changes in different gravitational fields. Rats have been grown for multiple generations under conditions of 2× gravity. The animals had shorter legs with increased bone density, but the increase in gravity did not appear to change development or behavior once they had adapted to their increased weight. Chickens have also been raised on centrifuges and had shorter, thicker legs. Based on these two examples, humans born and raised on the moon will be taller than their parents and possess more fragile bones.

Antigravity muscles and bones will have the greatest decrease in size and strength. There will be some decreases in overall bone and muscle development, but not as much as seen in our antigravity tissues. Can a prediction be made today, based on other information, of how much that decrease might be in second or third generation inhabitants on the moon? First, they will be born not adapted to their environment. During gestation they will be floating in their mothers' wombs, just as all mammals have done since they evolved millions of years ago. They will only experience the gravitational field where their mothers live. As they grow and mature, they will develop bones and antigravity muscles that are right for the gravitational fields of the moon or Mars.

How might this change their structures? One prediction can be made based on the relationship of percentage of bone and body weight in Earth's gravity (Figure 10.2). An elephant with a body weight of about 7,000 kg has a bone mass of approximately 27%. A man with a body weight of 70 kg has a bone mass of about 15%. A 10 kg dog is 10% bone while a 20 g mouse is 5% bone. If the 70 kg man went to the moon, he would weigh 70 times the moon's gravitational field or 70 × 0.16, equaling 11.2 kg. Once adapted to the lunar environment, his body weight would perhaps decrease to 10 kg based on disuse atrophy of antigravity muscles. It seems a reasonable prediction, at this time, that overall he would be about 10–11% bone instead of his usual 15%. Children born on the moon would definitely have a decrease in their bone structure. The same approach can be used to estimate bone mass on Mars or other gravitational fields.

It is more difficult to predict other changes likely to be seen in long term residents or subsequent generations. Yes, antigravity muscles will be smaller, but body muscle mass will continue to be dependent upon genetic background and type of behavior.

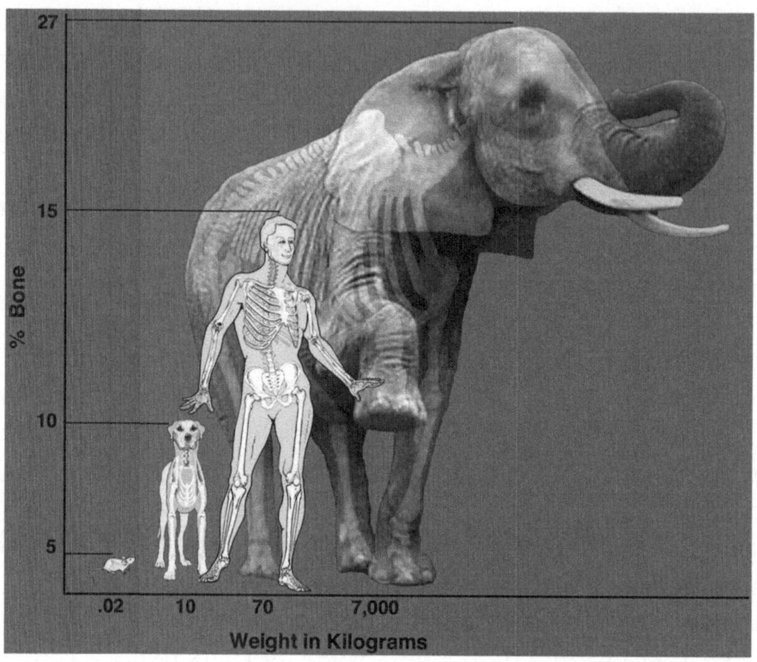

FIGURE 10.2 On Earth there is a direct correlation between body weight and percent of the body that is bone. Information such as this can be used to make predictions regarding human bone mass in individuals who reside for long periods, or mature, in alternative gravitational fields, such as the moon or Mars, where their weight will be less than on Earth.

Body builders who pump iron on the moon and Mars will have an increase in muscle bulk and strength just as they do on Earth.

Since we know without question that animals, not only can but will adapt to μg, we can be assured that they will also adapt to intermediate gravity fields. Regarding adult humans, what can we expect? Bone structure will be decreased. The changes will be greatest in those parts of the skeleton that were shown to be most affected in μg, those were, the pelvis and legs as well as the lower lumbar vertebrae. Non-weight bearing bones like the skull, ribs, and arms have little change in μg. It is unreasonable to expect that they will be significantly changed on other planetary surfaces. Will it make a difference if organisms move to a new world in their developmental stages? Or perhaps are born and complete their life in this new environment? There may not be a great deal of anatomical effect in such an individual's non-weight bearing bone or associated muscles, but that is conjecture. Changes that occur after generations in decreased gravity will be greatest in the systems that allow us to defy gravity on Earth.

It is obvious from the Apollo flights to the moon that the astronauts who arrived for a short stay on the surface had changes in their normal gait and mobility when outside the lunar lander. Unfortunately, they were not on the moon long enough to have the opportunity to adapt. Learning to walk was confined to a few brief trips upon the lunar surface. Space inside the lander was quite limited, about 6.5 m^3. A room that size with a short 7 ft ceiling would have a 5 ft wide wall on each side. That is much too small a space to allow the astronauts to practice walking, much less running or jumping. When on surface excursions away from the lander, they had to wear awkward and cumbersome lunar space suits. The lunar space suits weighed about 180 lb on Earth and would have weighed 30 lb on the moon. The total weight of an Apollo astronaut and suit was about 55–60 lb on the moon. Their movements about the surface appeared as a combination of a hop and a shuffle. moon walkers had difficulty picking up dropped items and stumbled easily. In the slight gravity, after falling, they were able to bounce upright in spite of the space suits. They moved freely about the lunar terrain. See Video #013: Apollo Astronauts on the moon.

It will be interesting to watch lunar explorers in the future, after they have had an opportunity to adapt. Movements of personnel inside a large lunar habitat or colony facility, where they do not need to wear space suits, would be astounding from an Earth perspective. Before extensor muscles and supportive bones significantly adapted to the new environment, a person's physical ability would greatly exceed anything that could be accomplished on Earth. Prodigious leaps and spectacular aerial aerobatics would be commonplace if the habitat were sufficiently large to allow it. To add to the ability to jump higher because of the decrease in body weight, there will be a much slower rate of fall, due to the decreased gravity. The speed of free fall acceleration in the moon's gravitational field would be only 1.6 m/s/s. Really a slow drift down compared to the 9.8 m/s/s that is present on Earth. This slow fall and long hang time will greatly increase a gymnast's ability to show off his/her skills.

During adaptation to gravity on other planetary surfaces, individuals would slowly adjust their walking habits to deal with the new conditions. moon dwellers will not shuffle, hop, and stumble as frequently after they have been on the lunar surface long enough to become adapted to 0.16g. They will need to be dual adapted, wearing an excursion pressurized space suit when out on the moon's surface and enjoying the freedom of a shirt sleeve environment in the habitat. Consistent with alterations in movement, there will be a change in neuromuscular response.

Imagine a ball being thrown through the air inside a lunar habitat. Its forward velocity would be a bit faster than seen on Earth as the atmospheric pressure or air density will be decreased. There will be a lesser percentage of nitrogen and an increased oxygen percentage with a decreased total pressure. One reason to decrease total pressure is that less nitrogen will have to be imported to the moon where it is essentially non-existent at this time. The gravitational pull on a thrown ball is much less and it would fall more slowly to the surface. Lunar dwellers engaged in a game of toss and catch will rapidly develop visual/neuromuscular changes to anticipate the new path of thrown objects. If the same ball were thrown outside by an astronaut, an even greater throw length would be possible if the astronaut could overcome the limiting features of the space suit. Without an atmosphere, there is nothing to slow the ball until the moon's gravity pulls it down to the surface. The ball would go farther in the same length of time. During the Apollo 14 mission astronaut Alan Shepard hit several golf balls a considerable distance on the moon's surface with a makeshift golfclub.

For the mathematically inclined with a couple of assumptions, you can compute an anticipated distance that a ball might fly without an atmosphere to slow it down and a 0.16 gravitational field. Two major assumptions would be the effect of no atmosphere slowing the ball's flight and the strength of the golfer's swing while wearing a lunar surface suit. The same sort of contest could be arranged with a football or baseball and bat.

A problem with a long stay on the moon will become evident when the lunar-born wants to visit or immigrate back to Earth. With decreased bone and muscle strength in the antigravity systems of his/her body, strenuous physical activity will not be possible when you are first exposed to Earth's gravitational field. There is the danger of hip and leg fractures, and there would be an increased risk of muscle damage as individuals accommodate to the unaccustomed work of struggling against Earth's gravity.

Since it takes 2.5 times as long to regain bone as it does to lose it after stays in LEO, rehabilitation to Earth gravity will be a long and tedious process for long term lunar dwellers. Numerous experimental animal multi-generational studies will need to be conducted prior to exposing humans to a lifetime of low gravity. Given the rapid generation time of most laboratory animals, this is not an impossible task.

Re-adaptive changes on Earth, following a long stay on Mars, would be similar to the problems encountered after returning to Earth from the moon, but not as serious. With the length of travel time to get to Mars and the need to synchronize Earth-Mars positions for transit, it is likely that stays there will be lengthy even in the early years of Martian exploration and development.

Part B: Destination M & M: No Motels, No Fast Food

Feeling my womb o'er pregnant with the seed of cities unborn
Wild and wide are my borders, stern as death is my sway,
And I wait for the men who will win me—and I will not be won in a day;
And I will not be won by weaklings, subtle, suave and mild,
But by men with the hearts of vikings, and the simple faith of a child;
Desperate, strong and resistless, unthrottled by fear or defeat,
Them will I gild with my treasure, them will I glut with my meat.

Robert W. Service (1874–1958), Poet,
from *The Law of the Yukon*

The moon and Mars are clearly in our sights as a first step away from Earth. Initial attempts at living away from Earth are in our future. First the moon and then Mars will be explored and settled. The first movement to new worlds will be by twenty-first century pioneers and explorers. Based on their experiences of living away from Earth for periods of time in relatively simple habitats, we will progress to outposts where there will be extended stays. These early ventures will be followed by permanent colonies with lifelong and multi-generational residency. With that experience, we may indeed move out to inhabit other portions of our solar system, perhaps eventually beyond. This will be when Earth-like planets are identified around other stars.

Other factors need to be considered in making plans for extended trips away from Earth. Living in a tightly sealed spacecraft, habitat, or colony for prolonged periods, months to years, raises concerns that potentially toxic substances may slowly increase in the atmosphere causing either acute or chronic sickness. In the early days of coal mining, similar problems occurred down in the mines. As a solution, miners carried a canary in a cage when they went into the mine. As long as the canary was healthy, the miners did not worry. However, a sick or dying canary was a signal to leave the mine and get into fresh air. Today we have more sophisticated ways to detect toxic perhaps odorless gases that may be dangerous to space travelers. One of the best is called an electronic nose. A current model of the electronic nose is in use on the ISS at this time. In the future, more advanced models will be used on long voyages and in extraterrestrial facilities that of necessity are closed systems. The atmosphere must be continually recycled and replenished to support all of the Earth life: humans, other animals, and plants. In this kind of living and working environment, it is quite likely that introduction of small amounts of potentially toxic substances into the atmosphere will occur.

The job of the electronic nose will be to detect the harmful compounds and to alert the air quality control officer before the levels become dangerously high. The nose can be constructed to detect and identify many different and potentially dangerous airborne substances. At this time, versions of the electronic nose are quite small, about the size of a large paperback book. We have come a long way from a canary in a cage.

To begin living on the moon or Mars, the first consideration needs to be the resources present. A requirement to transport large quantities of material from Earth would doom any hope of permanent living. Without the ability to develop a self- sufficient infrastructure, colonies will be short lived or never initiated. To be successful, it will be necessary to 'live off the land.' That is, of course, a gross oversimplification. The land is there, but not an atmosphere that can support Earth life. Water may initially be in short supply. We will need to build enclosed facilities/habitats and continually recycle, reuse, recycle. Nothing that has to be imported from Earth can be discarded. It will be a new life style for our often wasteful society.

There are many resources available on Mars, but less so on the moon. In a way that is acceptable, introducing or supplying some raw materials to our moon is much more feasible than exporting them to Mars. Mars is over 100 times farther away at its closest, and that is only for a few months at a time. The rate of rotation around the sun by Earth and Mars is too different to have ready re-supply or help available on an 'as needed' basis. Much less energy and money will be expended per pound of payload delivered to the moon. Nevertheless, Mars has better conditions to support a settlement. We know that there is water present and near to the surface. As we begin to expand to the moon, living there for a few years will be a big help for later establishing living facilities on Mars. One reason for developing a colony on the moon is that by doing so, we will acquire the knowledge and skills that will allow us to make the next move to the red planet.

To be self-sustaining means that we will be required to construct our own housing, food production, engineering, and research facilities. In addition, it will be necessary for us to move about the surface on foot and in self-propelled vehicles. All of these capabilities must be present. Most raw materials needed for construction are known to be present on both the moon and Mars. Still, the colonies must be jump started as the necessary tools and equipment to turn raw materials into living and working facilities are not on-site ready for the settlers to use. We must prime the pump with equipment, essential goods, and services.

Let's compare Earth, Mars, and the moon to see what they offer as homes for humanity. Figure 10.3 presents an overall perspective of these three unique and different locales. We know that people must live inside pressurized facilities and wear temperature controlled, pressurized spacesuits when outside on either the moon or Mars. One psychological or perhaps visual factor will be important for most of us as we acclimate to our new home on these strange worlds. There will be no open water anywhere on the entire landscape! That will be a new experience. To adjust to the complete lack of natural streams, lakes, or oceans, no rain and no snow will be a downside to living away from Earth. The opportunity to gaze out at water, running or stationary, is important to our psyches. This will be hardest for the original immigrants from Earth. For our children, born and raised away from Earth, what they grow up with will be the natural or normal world. Being in the middle of a featureless expanse of water will be a frightening experience to lunar or Martian natives born and raised on these other worlds if they venture back to Earth and travel across an ocean or even a large lake. Can you imagine a lunar native shooting the rapids during the spring runoff in a mountain river? What a wild experience that would be!

In the beginning, everything needed by lunar explorers and scientists will be carried from Earth. Not all of this shipping will require the presence of human crews. It will be more expedient, and much cheaper, to send necessary cargo in transport vehicles without a human crew. They can be soft-landed on the surface where they will be used (Table 10.1).

One resource that isn't on the moon nor probably Mars is fossil fuels. Lunarians and Mars dwellers will have to learn to live without internal combustion machines that use petroleum products. Solar and nuclear systems may well be the norm for providing electrical power. While Mars has carbon, the moon does not have enough of this essential element needed for life. The carbon that is there was presumably deposited by the solar wind. It is present in the lunar regolith at about 200 parts per million (ppm). Even if you could extract 100% of that carbon, the extracting plant would need to process 2.5 t to recover just 1 pound of carbon. To put that in perspective, you can get that much carbon from burning a dry log that weighs only 3 pounds or importing 1 pound of carbon to the moon. Some carbon extraction could be a by-product of other mining activities in the future, but this will be of no immediate benefit.

The absence of petroleum or organic products means that many plastics cannot be synthesized. On Earth, 4% of the world's petroleum production is converted to plastic. Food packaging, clothes, and many structural materials like polyvinyl chloride (PVC), house siding and pipes, are an essential part of daily life in the United States and other countries. In much of the industrialized world, we have become dependent on a ready and inexpensive supply of easily discarded plastic.

Fortunately there is another option. A new biodegradable plastic industry has begun to develop that uses starch and cellulose-based polymers from plants to form plastics. Once an off-Earth colony has been developed, some waste plant materials could

	EARTH	MARS	MOON
Diameter (miles)	7900	4200	2160
Relative mass	1	0.1	0.012
Gravity	1	0.385	0.16
Mean surface temperature (C)	15	-55	-15
Atmospheric Pressure mmHg	760	6	negligible
pounds/square inch	14.7	0.12	negligible
Atmospheric Composition %			
Nitrogen	78	2.7	-------
Oxygen	21	--------	-------
Carbon dioxide	0.04	95.2	-------
Escape Velocity miles/hour**	25,350	11,250	5,000
Surface	29% land	100% land	100% land
Distance from Sun AU***	1	1.5	1

* Given the differences in diameter of the three bodies, the horizon will be much closer than Earthlings are used to when they arrive on the surface of Mars and even more so on the moon.

** Escape velocity is the speed required to escape a planetary body's gravity. It will be more economical to launch payloads from the moon because of the low escape velocity necessary. It has been proposed that it can be accomplished entirely by electromagnetic propulsion with no on-board fuel necessary to leave the moon's gravitational field. Adding to the ease of launching payloads from the moon is the lack of atmospheric drag.

***AU or Astronomical Unit is the mean distance from the Earth to the Sun. It is about 93,000,000 miles. It is commonly used to indicate relative distances within our solar system. It is too small to be useful as a measure of distance to other solar systems.

FIGURE 10.3 A comparison of characteristics of Earth, Mars, and the moon.

be used to manufacture plastics. Although not yet an industry, some plastics can be formed from carbon dioxide. Ethylene can be formed from carbon dioxide if hydrogen from the breakdown of water is available. Ethylene, in turn, can be converted to polyethylene and polypropylene. After the process of synthesizing plastics from CO_2 is developed on the moon, the same process could easily be used on Mars. It has a relative abundance of carbon dioxide in its atmosphere, as well as water underground.

That ever present dust covering the surfaces of both the moon and Mars will be hard on the moving parts of equipment imported from Earth or built on site. Spending many millions of dollars to import a special piece of equipment that has a very short functional life would be counterproductive. Specialized equipment must be designed to not only work in lunar or Martian gravity, but it must also withstand other harsh environmental factors that are present.

For us to leave Earth and start our new life on either the moon or Mars, we must utilize to the extent possible raw materials that are present at our new locations. The elements of greatest concern are oxygen, carbon as carbon dioxide or in organic molecules, hydrogen, and nitrogen. Water, H_2O, is an accessible source of both oxygen and hydrogen as long as conversion energy is available.

It has been proven in recent times that there is water at the south pole of the moon. It is frozen in deep crevices where the sun never shines. We suspect that it is also available at the moon's north pole. The big question now is how much? Water can be broken down into hydrogen and oxygen to be used for both propulsion and life support. Oxygen and water are requirements for multicellular life, plant and animal metabolic activities. The settling of the moon will be much easier if large stores of water are located and accessible to explorers and colonists.

Lunar development will serve as the training ground for exploration elsewhere in the Solar System and eventually beyond. The moon's proximity makes it an ideal location to begin to learn to explore other worlds. New equipment will be relatively easy to send to the moon and to test. The value of new technologies can be easily determined, then modified or discarded as appropriate. We are fortunate to have an accessible "testing grounds" so close to our home planet.

NASA has chosen the lunar south pole as the probable site for a first United States outpost on the moon. The exact location has not been chosen, but the Shackleton Crater is a strong candidate. It has sunlight over 70% of the time which will facilitate solar power generation. It is also close to recently located water deposits.

A problem with early designs for living on the moon is that the outpost may be a mobile home-like structure on the surface. There would be no protection from GCRs or micro-meteorites, not to mention larger falling objects that could destroy surface structures. For those planning to reside on the moon for longer periods or to immigrate to this new world, underground living quarters will need to be developed as a protection from hazards falling from the sky. Living permanently underground in enclosed pressurized facilities that include living and working facilities will be a new experience for most of us. Although not common, there are facilities like this in large metropolitan areas where individuals do not need to go outside unless they so choose. Everything is available in connected underground or high rise units.

Nitrogen is not present on the moon. A supply of these two elements, carbon and nitrogen, are needed to support life in an outpost or colony. They must be imported, conserved, and recycled so that large quantities may not need to be carried from Earth. Without an atmosphere on the moon, all Earth life will be contained in closed and pressurized environments. During the early missions, they will be simple chambers, but more spacious facilities will be constructed as human habitation grows and evolves. There will be some gas losses due to leakage, but recycling by both physical-chemical and biological means should allow for the development of robust outposts or colonies. However, a continuing source of biologically usable carbon and nitrogen will have to be supplied to lunar facilities from Earth.

A long range goal for lunar habitation is to develop a miniature closed Advanced Life Support (ALS) system that in many ways mimics Earth's biological environment. It will fulfill a number of important needs of the growing human population. The plant components will supply the need for raw materials to be turned into food, oxygen for the animal population to breathe, and clean pure water for them to drink. The plants will also keep the atmosphere usable by removing carbon dioxide. Finally, they will provide a remembrance of Earth's abundant plant population. A touch of home away from home. Both Earth and the moon are in position to use the sun to supply energy.

The great difference between these development plans for the moon and reality of Earth is that very small fluctuations in the atmosphere or temperature of the small lunar system, or breakdown of key components, could have major effects in a very short period of time. Except for occasional major collisions with other large objects in space that have devastated Earth life, our planet's ecological system is pretty stable. The Earth's system has large reservoirs of resources. Thus it is able to overcome or react slowly to the many changes brought about by our expanding populations. Unfortunately, sheer numbers of people are becoming overwhelming and our ecological system is not infinite. The current overabundance of our species on planet Earth is depleting our natural world. Unless population growth is reversed, collapses of major ecological systems may well occur.

Figure 10.4 is an artist's depiction of an outpost covered with containers filled with lunar regolith for radiation and micrometeorite protection as well as heat conservation. It has multiple levels serving different functions. It is compact but utilitarian.

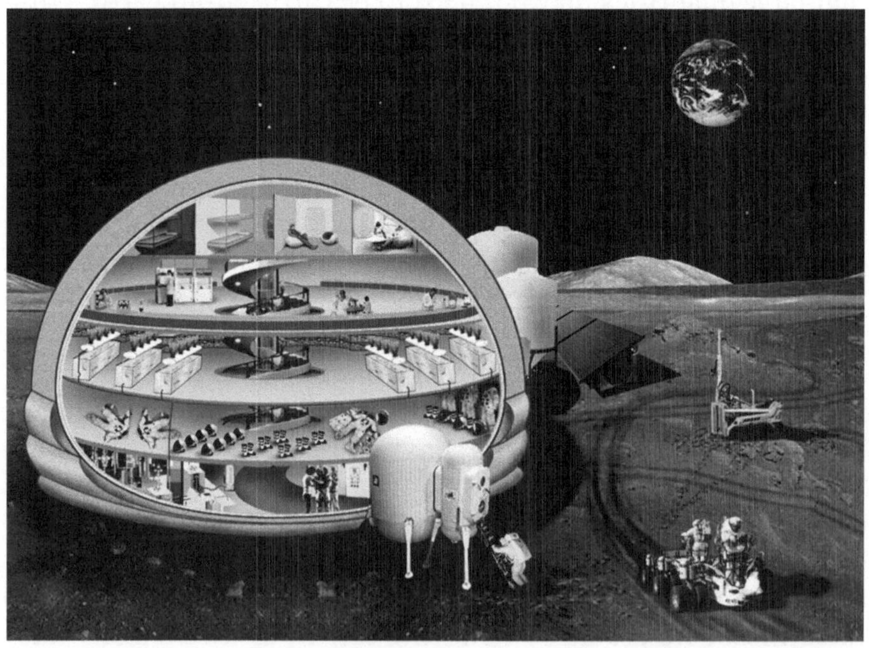

FIGURE 10.4 Artist's concept of a regolith covered habitat or outpost for use on the moon. Similar structures could be used on Mars.

Figure 10.5 is a more detailed view of the animal protein farm in the facility. Sections are provided for raising chickens, rabbits, and fish. Such a production unit would be integrated with food processing, meal preparation, and resource recycling components of the outpost. We sometimes forget or overlook the important step of food processing. In developed societies today, it is all accomplished in some far off and unknown facility. Early lunar settlers will be back to the stage of an agricultural community in earlier times. Food processing is just the next step after harvesting. Cereal grains, for example, will need to be turned into flour and then flour into bread.

Programs for space exploration and for establishing outposts and colonies will be a progression of developing facilities with new features and increased diversity. This will be true for both government and privately funded commercial operations. Early efforts will focus on well-known physical-chemical means for recycling ingredients for human use. Both oxygen to provide a breathable atmosphere and potable water will be essential. The basic

FIGURE 10.5 Artist's concept of a lunar facility for raising fish, chickens, and rabbits on one level of the habitat shown in Figure 10.4.

requirements of such a system will be to convert carbon dioxide exhaled by humans to oxygen that can be supplied to the pioneers. Waste water, in turn, must be re-purified before it can be used. Physical-chemical systems for life support are well established and have been essential components of long space flights. However, they do not save and recycle all of the carbon, just carbon dioxide. As long as all food is imported, recycling all carbon is not a necessity. Saving and stockpiling all imported carbon for future use should be an integral part of starting to live on the moon.

In the beginning all foods will be preprocessed, imported from Earth, and essentially ready to eat. Even at an early stage, it would be wise to store carbon rich wastes for later use in a biological recycling system. This would include biodegradable plastics as well as human and food wastes. The new pioneers will need to initiate biological recycling systems. Organic compounds cannot be wasted, but will need to be conserved to support the developing biological life support system that will come on line as the settlement matures.

There is no question learning to live away from Earth will be a long term effort. A colony cannot be considered as successful until it is self-sufficient. That is, growing foods that will support the inhabitants and recycling waste materials. Without supplying the nutritional needs of the personnel by growing and processing foods on site, the venture will be little more than a sophisticated camping trip. The NASA philosophy, at this time, is to primarily rely on physical-chemical life support to the exclusion of biological systems that would also provide nutrition. To that end, funding for Advanced Life Support development has been reduced by NASA.

The other possibility is that commercial colonization efforts will be more willing to embrace the inevitable and direct their efforts toward developing a closed ecological life support system. They may not have the built-in lethargy of governmental programs to avoid changing from the known to the unknown. The underlying suspicion of many engineers is that a biological approach is difficult and perhaps impossible to accurately quantify. This seems to make them hesitant to proceed in that direction.

The plans for early plant growth will be for esthetic and psychological reasons as well as the more obvious nutritional requirements of the crews. Based on both Antarctic experiences and long space station missions, the presence of "green growing things" has a very positive psychological effect on the crew's well-being. This point was well stated by Valentin Lebedev in his book *Diary of a Cosmonaut: 211 days in Space* (Bantam Books, 1995). In it he spoke of the aftermath of sending space plants back to Earth.

> During a TV broadcast we admitted that we feel sad and uncomfortable without our garden and without our dear plants. It was such a pleasure to take care of them. Man probably has a need to take care of things, and without those things feels empty.

Essentially the same conclusion was stated by another Russian cosmonaut on an early Soviet Salyut mission. He was speaking of flax, cabbage, and Hawkshead seedlings that they were trying to grow and said, "they are our love." The need to care for green growing things will be important, beginning with the first moon outposts or the initial trip to Mars. The first plants to be grown will have a double purpose. They will add interest to the meals as well as provide psychological support.

Once experience of early plant biological life support systems has been gained and proof of concept proven. It will be time for expansion to other plant crops that may need to be processed prior to utilization. Before taking that step, we need to consider lunar resources to support both an initial 'salad machine' and a more robust and extensive system capable of providing overall life support for the residents. One downside to living on the moon and beginning to establish a workable biological regenerating facility is the lack of an atmosphere. Here on Earth, we are accustomed to a thick atmosphere, about 760 mm of mercury at sea level (Figure 10.3).

One problem of living on the moon is the absence of nitrogen to construct an artificial Earth-like atmosphere. We know from the disastrous fire during a simulation in the Apollo 1 command module that 100% oxygen can be very dangerous. Skylab in the 1970s had an atmosphere that was 74% oxygen and 24% nitrogen with an atmospheric pressure of 260 mmHg, about one-third as great as that on Earth at sea level. No biomedical problems were associated with the decreased pressure or slight increase in oxygen availability on the three crewed missions to Skylab. Based on that success, it would seem that a comparable atmosphere would be strongly considered for a lunar base and future colony. In addition to requiring less nitrogen transport from Earth, the decreased pressure in the habitat, at only 260 mm of mercury, would result in a slower rate of gas loss, via leakage, from the lunar facilities than if it were at Earth normal pressure. Even though there is no native carbon dioxide on the moon, the requirement will be to only have a concentration of about 0.12% to be comparable to Earth carbon dioxide concentration in order to support plant growth. Actually, CO_2 concentration can increase substantially without detriment to plants or animals. On current space shuttle and ISS flights, carbon dioxide may increase to ten times the normal concentration in Earth's atmosphere. Such an increase actually enhances growth for many plant species. This is particularly true if the increased CO_2 is coupled with more intense lighting. Wheat grown under conditions of increased light and carbon dioxide produced a wheat crop three to four times the world record that was achieved with normal lighting and CO_2. A potato crop was over two times the world record. We need to find out, here on Earth, before we commit to outposts and colonies on the moon what an optimum CO_2 concentration and lighting level will be for plant growth in a decreased atmospheric pressure.

There are additional benefits to decreasing nitrogen in the lunar facility's atmosphere. The concern over developing the bends during excursions on the moon's surface would be greatly lessened by lowering the percentage of atmospheric nitrogen. This in turn would decrease the quantity of nitrogen dissolved in the blood stream of inhabitants when they prepare to go outside. The decrease in total pressure would also affect plants. With a decrease of this magnitude, the rate of transpiration by plants would increase as there would be an increase in atmospheric water vapor pressure; the tendency to evaporate. Plus less nitrogen would need to be imported from Earth.

As a side note, at this low atmospheric pressure, water begins to boil at about 65°C or 150°F. Such a change in boiling point will affect food preparation and recipes. All of our recipes are designed near a boiling point of 212°F. Foods will have to cook for a longer time in order to be ready to eat. According to some French chefs the proper temperature for cooking an egg is 155°F. Egg-eaters may be pleased with the lunar low temperature boiling point if it produces a more tasty product. Perhaps in lunar or Martian kitchens there will be a resurgence of pressure cookers, which are designed to cook at high pressures and high temperatures. In addition to the moon being a valuable testing location for learning to live away from Earth, there is an economic reason for establishing a lunar base and then a colony. The moon's regolith contains a relatively high concentration of the element helium 3. It has much more of this helium isotope than is present on Earth.

Using fusion technology, helium 3 can interact with deuterium, a non-radioactive form of hydrogen. In the process, a tremendous amount of usable energy is released. The final products of this reaction are helium 4, hydrogen, and energy. No carbon dioxide is released and no radioactivity is involved. Helium 3 is a new, environmentally clean source of energy (Figure 10.6).

Projections are that there are a million tons of helium 3 on the moon. Experts have stated that if 25 t of helium 3 were sent to Earth, when processed, it could supply the energy needs of the United States for a year. From another perspective, the moon's

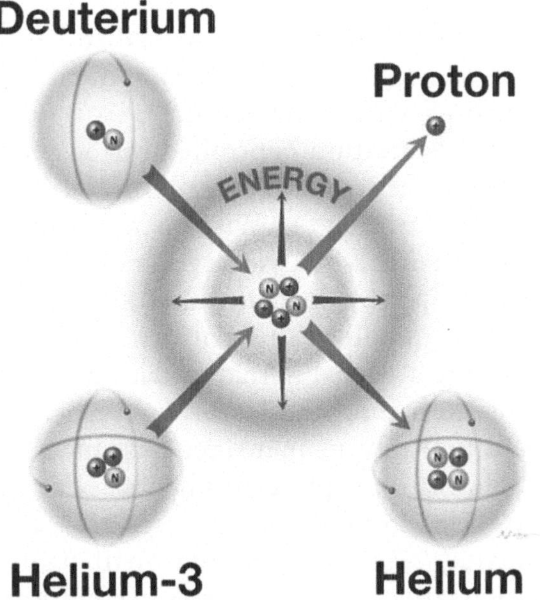

FIGURE 10.6 Using high temperature reaction chambers, it is possible to fuse helium 3 and deuterium nuclei.

stores of helium 3 represent ten times the total amount of fossil fuel oil, natural gas, and coal that has ever existed on Earth. We are at or past our peak ability to increase fossil fuel production, yet the demand for energy continues to rise. We are heading down a slope of decreasing availability. One or more new fuel sources must be developed or civilization as we know it will begin to fade away. Hopefully civilization will begin to learn that all resources are finite.

A new space race to initiate mining and shipping helium 3 to Earth is on. The slight lunar gravitational field will be a big help in this endeavor. The mined product, stored in large tanks, can be sent into space on a Earth trajectory using electromagnetic propulsion instead of big chemical rockets. Remember, escape velocity from the moon is only 5,000 mph as noted in Figure 10.3. With no atmospheric drag to overcome, that is an achievable speed. At this time the new space race involves at least 5 different countries or consortiums. They are Russia, China, India, Japan, and the European Space Agency (ESA), a consortium of European countries. In addition, several commercial entities are developing plans to go to the moon and initiate helium 3 mining. These countries and organizations anticipate developing an outpost on the moon on or before 2030. The collective goal is to establish helium 3 mining facilities as soon as they can. Robotic equipment will be used to do the mining and separating the helium 3.

The helium 3/deuterium controlled fusion process is under development around the world. A functioning laboratory model is in operation at the University of Wisconsin. The hope and desire is that the ability to reliably release large amounts of usable fusion energy will be developed by the time that lunar helium 3 becomes available around the middle of this century.

Mars represents a very different situation from the perspective of being able to live off the land. Water exists on the planet at the poles and probably in many regions underground, which will be accessible to the explorer/residents. The atmosphere on Mars, although slight from an Earth perspective, is composed of 95% carbon dioxide, 2.7% nitrogen, and small quantities of other gases. It will be a relatively simple matter to compress the Martian atmosphere, remove carbon dioxide, and add oxygen from the dissociation of water to create a breathable atmosphere inside Martian facilities. There will be no need to import gaseous raw materials or carbon in any form. The presence of these essential ingredients of life will greatly facilitate the establishment of a viable Martian community. Mars has a gravity field over twice as strong as the moon, so it will be easier for you to adapt. Bone and muscle loss will not only be less severe, but also less rapid than on either the moon or in μg.

One intriguing concept for a Mars habitat and living area, particularly in the early days of developing settlements, is currently being studied. It is to use some of the many large lava tubes that are present as a result of earlier volcanic activity. They are near Mars' extinct volcanoes. Martian lava tubes are believed to be much longer and larger than lava tubes formed near volcanoes on Earth; perhaps 50 yards or more across, easily large enough to develop a community (Figure 10.7). A lava tube/cave habitat would

FIGURE 10.7 Cross section of a Martian colony developed inside a portion of a long lava tube already present on Mars' surface. The tubes will provide a ready-made protective cover to shield from radiation and small meteorites. The design would place an inflatable structure in the tube, construct a lower physical planet and equipment level, and have the colony living areas on the second level. The colony could be enlarged by adding additional inflatable sections to other portions of the tube.

be protected from radiation and meteorites. It would be easy to maintain once established. Construction would also be easier as the tube provides the floor, ceiling, and exterior walls of the colony's structure. A section of the lava tube has been filled with an inflatable liner so that an atmosphere suitable for us can be maintained. The view shown is the second floor of the structure within the tube. Underneath would be much of the equipment needed to make the colony habitable. Atmosphere revitalization, storage, machine and repair shops, plastic synthesis, and other manufacturing processes would be conducted on the lower level.

Some lava tubes are miles in length and might ultimately be completely filled by one colony. There are individual, but separate, plant and animal growth units. An egress/ingress air lock with a parking garage is shown on the lower left. Other components, not all shown, would include living quarters, research laboratories, a large recreation area that would have community rooms including a library, television, exercise facility, computer room, and a swimming pool with a diving board to allow residents to capitalize on gymnastics in their low gravity home. Stores and shops would be added as appropriate.

Governmental offices, including engineering and design for colony maintenance and expansion, would be important, as well as medical facilities and schools. Transfer from one portion of the colony to another would be by bicycle, electric scooter, open electric cars, and trucks. There will not be high winds or inclement weather, no rain or snow, so closed vehicles will be unnecessary. Energy on Mars can be derived from nuclear reactors, geothermal sources, or thermal electric generators. Thermal electric systems operate by concentrating sunlight with a large mirror directed to a boiler. The heated fluid turns a generator to produce electricity.

To aid in Martian exploration, small airplanes with very large wings will be available for trips to distant areas. The planes will be constructed of lightweight plastic and will probably require that the pilot and passengers wear space suits. Even though the Martian atmosphere is slight, it is dense enough to allow travel by air.

Digging a number of meters below the surface to establish living and working quarters, Martian dwellers may come in contact with primitive life forms. Or we may find that Mars is barren.

There is no way to prevent the release of Earth organisms while moving about, on or under, the surface. Bacteria will be carried on astronaut suits and equipment wherever they go. Given the hardiness of some Earth bacteria, they may find a niche that they can occupy.

At this time, it cannot be predicted whether infants, born and developed on either the moon or Mars, would have skeletal, muscular, vestibular, or cardiovascular systems that would allow them to easily return to the native planet of their parents. Those questions must first be asked using experimental animals as human surrogates.

When the time comes to change from visiting to colonizing Mars, it will require public or private enthusiasm and a large resource commitment. Early exploration flights to other worlds will probably continue to be funded by individual governments or associations of nations.

The next step, from exploration to colonization, will likely bring a change from Earth governmental control. Local administrations, perhaps something akin to the "company towns" that were a part of the United States' progression from an agricultural to industrial economy in years past, may well develop. Those company towns were centered around a developing industry which may well be the case on the moon and Mars. This is appropriate as the role of government and their space agencies has been and should be exploration and discovery leading the way. As they are successful, they should stand aside and lead further explorations instead of attempting to be operational entities on our neighboring worlds. That should be the responsibility of the people who live there.

Part C: Hazards of the Moon and Mars

Of all of the hazards
Fear is the worst.

Sam Snead (1912–2002), Golfer

Before getting overly concerned with hazards on the moon and Mars, it is good to reflect on hazards here at home. Earth hazards that we have been exposed to since the beginning of time include: hurricanes, tornados, forest fires, floods, earthquakes, volcanic eruptions, carnivores, insect vector and parasitic diseases like malaria, bacterial diseases like bubonic plague or black death, scarlet fever, viral diseases like West Nile fever, measles, and poliomyelitis. More recent hazards include transportation in our modern society: cars, planes, and trains. Then there are wars, regional conflicts and vendettas that regularly kill large numbers of Earth's populace. Unfortunately, this last could be exported to other worlds. The difference is that we have been exposed to Earth hazards over countless centuries. These hazards are familiar and we tend to neglect them until they affect us personally. There has been no option of going elsewhere. Many of the hazards on Earth are regional and seasonal while on the moon and Mars they are continuous. Nevertheless, how do the new hazards that we know to be present on other planets stand up to our current and past hazards on Earth? Frankly, they are not too bad, even though they place limitations on our behavior. The most obvious restriction is that we must become used to living in enclosed, probably underground, communal habitats. It will not be possible to go outside without a protective space suit or pressurized vehicle.

Neither the moon nor Mars has a magnetosphere to block incoming solar radiation and GCR, so radiation exposure will be a constant problem, as already covered. The greatest danger will occur when one is outside with only a space suit for protection. Since GCRs come from all directions while on the surface, a person will be protected in one direction by the mass of the moon or planet but exposed in other directions. Due to the lack of an atmosphere, meteorites do not 'burn up' when approaching the surface as they do on Earth. Instead, they land intact and can damage living quarters or other facilities that are constructed without adequate protection. There are other hazards of concern. Both the moon and Mars have a loose and dusty surface that will cause problems for explorers. Although Mars has a slight atmosphere, it is not enough to support Earth life. The moon has essentially no atmosphere. Both experience extremes in temperature and, in particular, Mars is quite cold on the surface relative to Earth. Pressurized and temperature controlled space suits will be a requirement when one ventures outside.

The Moon

Without an atmosphere and never having had a period of time when liquid water was present, it is highly unlikely that life exists or has ever existed, on the moon. Biological hazards can thus be ruled out, including vector borne diseases. Based on the lunar material returned to Earth during the Apollo Moon explorations, there are no serious chemical hazards present in the regolith or rocks.

Physical hazards like radiation, extreme temperatures, and lack of an atmosphere are constant challenges of the lunar surface. Another problem exists. When individuals or equipment are outside, they will be exposed to the pervasiveness of the highly electrostatic dust that readily clings to space suits and equipment. During the Apollo missions, dust was carried into the lunar lander every time astronauts returned from exploration on the surface, even though they tried to brush it off and leave it outside. After exposure to moon dust inside the lunar lander, some of the remarks were "it's soft like snow yet strangely abrasive," "taste is not half bad," "it smells like spent gunpowder." There was at least one case of what we would call an allergy here on Earth. The affected astronaut had congested sinuses from breathing moon dust.

The problems with lunar dust are illustrated in Figure 10.8. It is a picture of a moon walker whose white space suit is covered with dirty gray dust. The next picture (Figure 10.9) is Gene Cernan after removing his space suit inside the lunar lander. It is obvious that the clinging dust particles coat everything with which they come in contact, including human bodies.

Moon dust forms when meteorites strike the surface, generating high heat that forms silica oxide, a form of sand, as a very fine dust. Tiny iron crumbs in these grains help to give them an electrostatic force (Figure 10.10). A benefit of having iron particles in the dust is that it will be possible to remove or separate the particles using magnets. This is one of the proposed solutions to prevent future lunar explorers from carrying lunar dust everywhere inside their moon base facilities. There are other mechanisms being considered for removing dust as astronauts return from exploration trips. One is to use an electrostatic sweeper in the air lock that one must traverse before entering the lunar habitat. Such solutions

FIGURE 10.8 Apollo 17, a picture of Harrison Schmitt on the moon's surface with a great deal of lunar dust adhering to his space suit. (Photo courtesy of NASA)

FIGURE 10.9 Apollo 17, Gene Cernan returned to the lunar module after an exploratory journey on the surface. The black smudges are tiny particles of lunar dust that were transferred from his space suit to his clothes and face. (Photo courtesy of NASA)

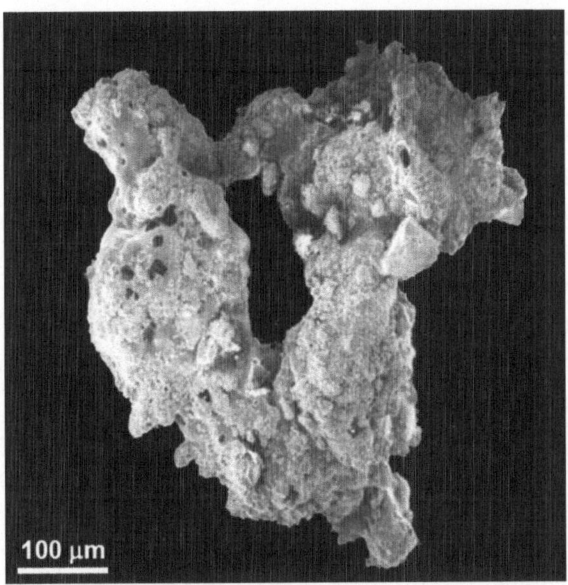

Figure 10.10 Photograph of a lunar dust particle. The very small iron fragments can be seen as dark spots on the surface of the particle. This dust is not only electrostatic clinging to surfaces but also can be very abrasive. This ever present dust may create a problem with bearings and moving parts of equipment imported from Earth or developed on the moon. (Photo courtesy of NASA)

could greatly lessen or even eliminate the problem of moon dust invading every nook and cranny of future lunar living areas.

Mars

From a hazard standpoint, Mars represents a different environment than the moon. There is the unknown possibility of Martian life near the surface or possibly even further underground where water is present. Deep underground, the temperature rises and reaches a level that could be favorable to Earth life. Liquid water and warm temperatures would be very conducive to supporting life. Based on the extreme environments where we find bacterial life on Earth, such as 120,000 year old ice in Greenland and organisms capable of multiplying at temperatures well below freezing in arctic seas, the subsurface of Mars may well contain a variety of living organisms. It is possible that initial permanent habitats on both the moon and Mars will be buried underground or constructed in caves or lava tubes.

From a chemical perspective, the surface soil of Mars is primarily iron oxide with the capacity to oxidize other substances that it contacts. On Earth, we are not used to such highly oxidizing soils. By convention, material on the surface of the moon is regolith. On Mars, it is designated as soil.

The potential physical problems that might be encountered on Mars may be different from those on the moon. The Martian gravitational field is a little over twice as strong. Even fine dust will settle more rapidly and remain on the surface. The electrostatic adhesiveness of Martian dust is not known at this time, but it could present a problem.

The Martian atmosphere is about 1% as dense as Earth's atmosphere. This atmosphere, as slight as it is, presents a hazard because Mars is subject to immense dust storms that can obliterate the surface for weeks at a time. The left portion of Figure 10.11 is a normal telescopic view of Mars. On the right is the same view with the planet engulfed in a global dust storm. Even though winds may exceed 100 mph, they are not destructive in the sense of a

FIGURE 10.11 Comparison pictures of Mars by the Hubble telescope about 2.5 months apart in 2001. A severe seemingly worldwide dust storm has obscured the telescopic view of the planet's surface. The dust storms may last for weeks and be more severe in different regions. (Photo courtesy of NASA and Hubble Space Telescope)

category 2 or 3 hurricane of equal wind velocity on Earth. The slight atmosphere on Mars would make pressure from a 100 mph wind on an individual or a structure more like a breeze of 10–12 mph. The problem is not that of being blown over, but perhaps being blinded by dust/sand clouds. Currently, we have no first hand information regarding loss of visibility on the surface. Anything exposed on the surface will be covered with a layer of iron oxide dust and sand that also gives Mars its designation as the red planet. Driving a Mars exploration vehicle during such a storm could be hazardous for the driver as it may not be possible to see obstacles, such as boulders or even sharp drop offs.

Being "confined to quarters" for long periods, waiting for the termination of a dust storm would be worse than the similar situation in Antarctica during winter storms, as there will be no basis for judging when the weather will clear. This is bound to increase the feeling of isolation and the loss of control of your destiny.

Part D: New World

What place is this?
Where are we now?

Carl Sandburg (1878–1967), Writer,
Editor, Poet from *Grass*

Futurists have dreamed for many years of the possibility of constructing a new miniature world in the sky. A self contained city. The most common structure envisioned is a torus, a large doughnut shaped circular tube (Figure 10.12). As the torus spins, it creates a gravitational field that will be strongest along the inner surface of the outer circumference. The strength of the gravitational field will be dependent on the diameter of the torus and the speed of rotation. To the new residents, the direction down will always be towards the outer surface of the rim. As your great-grandchildren's descendants move from the outer ring towards the center, the gravitational field will decrease, but the feeling of rotation will increase. When they first arrive on this rotating world, they will have to adapt their vestibular systems to the rotation to avoid motion sickness. Most experts feel that if the structure is sufficiently large, it can spin at about 2 rpm and motion sickness

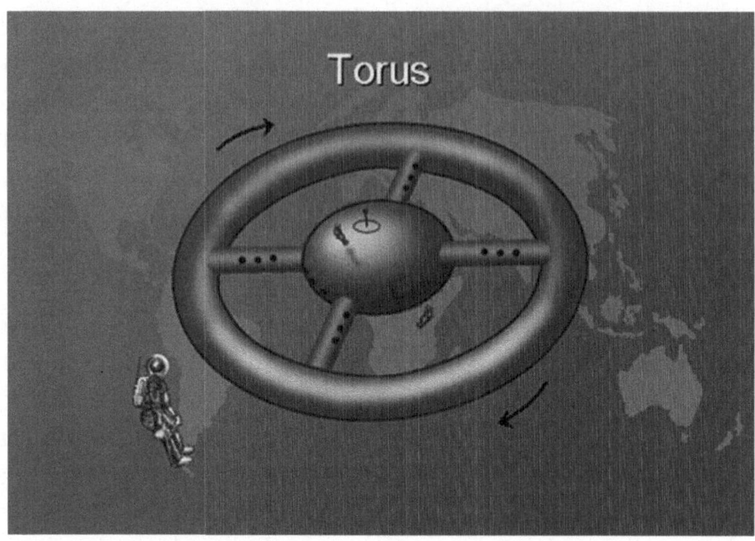

FIGURE 10.12 A torus envisioned as a space colony for a self-supporting miniature world.

will be only a minor concern. It is likely that as long as the rotation rate is no more than 7 rpm, inhabitants will be able to adapt to the continuous movement. By the time that such a structure is built, we will have had a great deal of experience with the moon's gravitational field and may design and rotate the structure to mimic gravity on the moon or Mars.

Our experiences living on the moon and on Mars will provide the knowledge and expertise to start this new habitat. Crops and animals already in use away from Earth can be imported. The technologies developed to recycle will be ready to support the operation of this fresh start for our descendants.

Where should we build this New World? It needs to be in a stable region so that it does not crash into Earth or the moon. There are several possibilities around planet Earth, fixed locations in the Earth-moon section of our Solar System. These are small regions where the gravitational pull from the moon and Earth are equal and an object in that region is in a stable area. Gravity exists, but the accelerations from the closest planetary bodies are equal and from different directions, so they appear to cancel each other. They are called Lagrange (L) points, where the gravitational pull

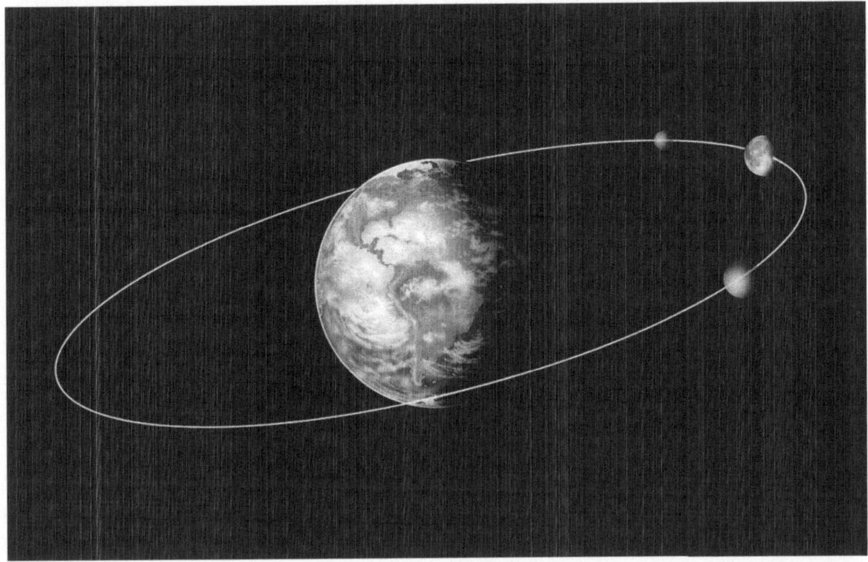

FIGURE 10.13 The most stable regions in space near our planet are those where the gravitational pull from the Earth and moon are equal.

of two planetary bodies are equal, forming a stable gravity region between them (Figure 10.13). L-1 is directly between the Earth and moon, but much closer to the moon than it is to the Earth because of the difference in mass of the two bodies. L-4 and L-5 seem to be the most appropriate regions. They are in the same orbit as the moon, but equidistant ahead and behind, and closer to the moon than to the Earth. They are the most stable of the Lagrange points and the likely sites for the future construction of a permanent rotating space colony. Colonies, once positioned, would remain in this stable location with only minor position shifts.

Living in space on a world that we have built is an intriguing concept. To do this, we will first need manufacturing facilities on the moon and simple launch systems from that location. Perhaps electromagnetic, or even anti-gravity, launch will be available. Before a "new world" can be seriously considered, we need to learn to live away from home on the moon and Mars.

The components that are needed for construction can be manufactured on the moon, launched by electromagnetic propulsion, and assembled at either L4 or L5. Alternatively, assembly could take place in lunar orbit and the completed structure slowly moved to the more stable L4 or L5 position.

The inhabitants of such a world with no natural resources must recycle and reuse everything on a continuous basis. They would need to produce some unique products to provide income and support for their space city; perhaps a nearby μg processing plant that did not rotate or perhaps knowledge, a University in space. Given the immaturity of the space program today and the cost of launching mass of any kind into LEO, much less to a stable Lagrange point near the moon, it is unlikely that such a structure as the New World will be built in the foreseeable future. We must first learn to crawl with regard to living away from Earth on already present worlds like our moon or Mars before serious consideration can be given to constructing an entirely new world, where the only natural resource will be the ingenuity of its inhabitants.

11. Taking the Plow to Mars

It is change, continuing change, inevitable change, that is the dominant factor in society today. No sensible decision can be made any longer without taking into account not only the world as it is, but the world as it will be.

Isaac Asimov (1920–1992)

We expect to take our food and supplies with us on a camping trip, maybe catch dinner on a fishing trip, or perhaps live off the land as the Lewis and Clark expedition did. However, this will not work if we really leave our planetary home. There is no life on the moon. If there is life on Mars, it is probably microscopic and not available for dinner. We are faced with the problem, particularly for Mars exploration and permanent habitation, to carry everything with us or to produce food and a breathable atmosphere from materials on hand. In the short term, we can probably supply a moon outpost with food, but as it becomes a permanent settlement, this approach will no longer be feasible. So, what to do?

For humans to begin to explore Mars the current plan is to carry all necessary foods and supplies on the initial trip. One of the approaches that has been considered is to send a habitat with laboratory and supplies to Mars before launching the crew. When that facility reports a successful landing back to Earth, final plans will be made to send the crew. If this plan is the chosen design, then the crew's food will have to be stable and nutritious for a long time. Landing on Mars with a crew of an unknown size to conduct intensive explorations will require a complex logistics system. Water will be recycled on the flight to Mars.

Depending upon success in finding exploitable water under the surface, it may be necessary to recycle Earth water while the crew is in residence. Although a ready water supply is essential, the more complex problem is providing food to the crew for a trip to Mars and back that would last over 2 years. A number of space

R.W. Phillips, *Grappling with Gravity: How Will Life Adapt to Living in Space?*, Astronomers' Universe, DOI 10.1007/978-1-4419-6899-9_11,

travel enthusiasts feel that the long voyage to and from Mars can be substantially shortened, but their approaches are still on the drawing board. For the present, the round trip to Mars will be long and tedious. Keep in mind Mars and Earth are over 40,000,000 miles apart at the closest. Mars goes around the Sun in 687 Earth days, so at times they are on opposite sides of the Sun.

There are two very different approaches being considered to supply a nutritious diet to explorer/astronauts on long range missions. Each plan has its advocates. One is to carry all supplies needed, including foodstuffs for use during the flight to/from Mars and while on the surface. This approach is the choice of many engineers because it fits with their experience supplying life support in LEO or on short moon missions. It is the physical-chemical solution, where oxygen and potable water are provided by machines that utilize well-developed technologies to change carbon dioxide to oxygen and render all used water potable. All food will be prepared and packaged on Earth prior to the mission. There is good evidence that the physical-chemical approach can do the job and provide water and nutritious food to the Martian explorers. But the physical-chemical approach is a dead end in the long run. It will provide a breathable atmosphere and water for explorers' use, but dooms space travelers to forever import all food. We need to learn to be self-sufficient so that we can begin to live on Mars as well as the moon.

The second approach, currently called Advanced Life-Support (ALS), is to create a miniature ecosystem capable of operating on other worlds. This approach is the one favored by life scientists. The plant and animal life on Earth have a true symbiosis. We animals utilize oxygen and produce carbon dioxide. We also convert reasonably clean rain and snow water into gray water, urine, water vapor, and other bodily wastes. Our companions, the plants, do the reverse. By the process of photosynthesis using light, plants convert carbon dioxide and water into oxygen and plant growth. We then use plant carbohydrates, lipids, and proteins for food.

We have discussed using crew carbon dioxide and air movement to stimulate photosynthesis and plant growth. By so doing, oxygen is provided for the crew to breathe. There is another valuable by-product of plant growth. It is pure water. We are familiar with the importance of rain for field crops and the necessity of watering house plants to keep them alive and growing. That is one side of the picture. It is also possible to obtain clear, clean water from plants in addition to food and oxygen. What a triple bonus that is! The drinkable water is due to a process called transpiration. When plants transpire, they absorb water from the soil and circulate it from the roots up to the leaves where part of the water is released as water vapor.

In enclosed plant growth units, like a greenhouse on the moon or Mars, it will be easy to collect that transpired water. All that is required is to install refrigerator coils and circulate greenhouse air saturated with water past the coils. Water will condense and can easily be collected. This is a much more efficient method of providing potable water for the crew than using the complex physical-chemical method of water purification.

Few people appreciate the quantity of water that is produced by plants. The actual quantity produced varies depending on the plant species, stage of growth, temperature, lighting, nutrient and water availability and growth space, but the message is clear. Large quantities of potable water can be condensed and collected during plant growth. Some is from evaporation but the great majority is the result of plant transpiration. In a functioning closed Advanced Life Support system, it is more than enough to satisfy the crew's needs for potable water. The nice thing about condensing transpired water is that it is just water without impurities, ready to be used for drinking, cooking, and hygiene, without further treatment.

Table 11.1 is a summary of information from the intensive growth of wheat, soybeans, lettuce, and potatoes, Irish and sweet, in the Biomass Production Chamber at the Kennedy Space Center over a number of years. This large chamber was used to collect information to aid in the future development of an ALS system. Edible plant mass was used as the basis for comparing various aspects of plant growth. What this table means is that for every pound of edible plant mass that was harvested, 473 pounds of usable water was collected. This is a real bargain, more than enough to supply the water needs in an outpost or colony. We can let the plants handle the job of delivering all of the water that we need, without engineering complex purification systems.

TABLE 11.1 Production and utilization by plants per pound[a]

Edible plant mass	1.0
Total plant mass	2.4
Oxygen produced	2.7
Carbon dioxide utilized	3.7
Water condensed[b]	473.0

[a] Nutrient content of the food will depend on the particular crop. Lettuce is high, while cereal grains, wheat and rice, have more inedible parts from our dietary perspective. Some other animals in the ecosystem can utilize straw and other plant wastes
[b] Water condensed in closed plant growth units comes from both transpiration and evaporation. Transpiration is by far the major component

An important question needs to be answered by competent, hopefully unbiased, experts. How long does a mission need to be to justify switching from a physical-chemical system to an ALS system? An ALS will grow the foods that the astronauts eat in addition to providing oxygen and potable water for human health and well- being. Some group, other than just NASA or industry engineers on one side and life scientists on the other, needs to be involved in making this important decision. There are so many variables that it is not easy to determine the right answer.

One way to help resolve this question is by using what has been called Equivalent System Mass (ESM) to determine the overall most efficient and least expensive methods to meet the needs of astronauts on long missions or to establish outposts on other worlds. ESM includes consideration of all of the factors that impact our ability to develop biospheres away from Earth that mimic Earth's biosphere. To enter into the development of an ALS system means that recycling inedible plant and animal products, as well as human wastes, will be standard operational procedure. The ALS will not only supply edible plant materials that can be processed into diets to sustain the voyagers, but ALS will supply the oxygen and pure water needed for human health and well-being. The system must be dependable, with a number of separate production units. If there was a single ALS unit and it suffered a serious breakdown, the results could be catastrophic. Redundancy will be a necessity for an ALS system and also for a physical-chemical system. This is particularly true for a Martian outpost or colony. Help is just too far away.

Developing and then operating a completely closed ALS system will require significant construction, maintenance, and crew time. Conversely, using a physical-chemical system, all food must be carried, ready to eat or ready to be prepared for eating, at the time of launch. The dietary components must be stable and capable of retaining quality following long periods, years, of storage. In addition, a physical-chemical system will also require redundant facilities and crew time. There is "no free lunch" in establishing permanent living quarters away from our planet.

We must first gain experience by developing a starter ALS on the moon and expand it until we are sure that we can build and maintain a full scale, closed ecological system on Mars or a New World. It must be a system that can nurture the various plants and animals and the human crew. Until its functionality is clearly established, there should be a combination of physical-chemical and ALS systems designed to provide both redundancy and dependability. This approach should be an integral component of a growing population of settlers on the moon. With such expertise as a background, ALS systems will be ready for development and use on Mars. Perhaps a physical-chemical system with one or more "salad machine" elements would be an appropriate initial step on the moon to gain experience. An ALS system can be very small and simple or large and complex, depending on the degree of closure. A system that produced all of the food, water, and oxygen needed would be 100% closed. A simple experimental ALS might be only 5% closed, perhaps raising lettuce and other salad ingredients.

There is another important benefit of ALS food production. The cost of sending seeds rather than edible foods to far distant locations is insignificant. Think in terms of a few pounds, instead of tons. Relatively few seeds will allow the growth and the development of enough crops to supply an outpost away from Earth. Preparing, packaging, and transporting a variety of very stable foods to provide nutritional support for a group of explorers for the duration of a Mars mission will require many tons of food. A physical-chemical system is appropriate for short missions but cannot support "living" on other worlds. To "live" away from

Earth requires self-sufficiency. We have a long way to go to reach that point. The major roadblock is the decision to get started.

ALS systems may be open or closed. Open systems are easier to establish and maintain. Constituents in excess can be removed, or if deficient, added. All that is needed is a sufficient supply of the necessary gases: oxygen, nitrogen, and carbon dioxide, as well as plentiful water. Major efforts are being made at this time to determine the availability of water on the moon. We know there is some water there, particularly in the polar regions but availability and quantity are unknown.

In creating a closed system, a balance must be present between oxygen and carbon dioxide concentrations, potable water, and crew nutrition. For example, the oxygen produced by photosynthesis by the plants must equal the oxygen needed by the crew and other animals. Animal oxygen utilization will be relatively constant, but oxygen production by plants varies by the amount of photosynthetic light provided and the stage of growth of the plants. New or very young plantings do not have enough leaf surface to produce large quantities of oxygen nor do some mature crops. A functional closed ALS system must be large enough to include plants in all stages of their life cycle for continuous production of food and oxygen.

Mushrooms can also be grown in small intensive horticulture systems. Edible mushrooms readily degrade cellulose, a plant structural product that cannot be directly used by humans. A small mushroom "farm" can convert inedible plant components into an edible delicacy. Other fungi like white rot can degrade lignins and make them available for use by other organisms. Lignin provides structure to plants and fiber in our diets when we eat vegetables. Most lignin comes from the woody part of plants. They make vegetables crunchy. The problem with lignin in the diet is that it is not broken down by our digestive systems. The fungi can make the lignin material available for use by humans and other animals.

Will the decreased atmospheric pressure in the habitats have an effect on plant growth? Is there a downside to making the facilities better for us? A number of scientists in the United States, Japan, Europe and Russia have studied aspects of this proposed atmospheric modification. They have found that plants are quite capable of growing and prospering in chambers with a decreased atmospheric pressure. Plus, there are all those other advantages; less gas to ship from Earth, thus less expense, less problem with the bends when you venture outside, and the opportunity to increase the carbon dioxide pressure in the ALS units to enhance the rate of plant growth for many species. It is unlikely that permanent colonies, or even outposts on other worlds, will try to maintain an Earth-like atmosphere in the working living quarters. Based on our experiences on Skylab, no problems are likely to develop while living in a reduced atmospheric pressure. It just does not make sense to provide an atmosphere that is 80% filler. It will be absolutely essential that there is enough nitrogen so that the atmosphere won't be combustible, but that is easy to do.

At this time, a biological revolution is occurring. We now have the knowledge and tools, using the latest in DNA technology, to genetically modify organisms, plant and animal. The hereditary material in chosen species can be changed to make it easier for them to adapt to conditions in lunar or Martian habitats. These efforts can be started on Earth and continue to be refined inside a lunar development. When the time comes, members of specific plant and animal species will be genetically modified for life away from Earth and ready to support the settling of both the moon and Mars.

Genetically modified (GM) plants designed for living off-Earth will not have some of the potential problems that have raised controversy over the past few years. The ability to produce pesticides will not be needed. There will not be insects (unless introduced) or other pests to contend with away from Earth. If such species were unintentionally exported from Earth to our new settlements, they would be very easy to control in the intensively cultured growth units. Individual units could be sterilized, as needed, if some disease or pest sneaks in.

Inserted genes could help plants more easily adapt to the new environment. Further, plants can be programmed to supply nutrients, like vitamins or essential oils, that may be in short supply in the limited plant species available.

One example is the relatively new, yellow rice that contains the important Vitamin A precursor beta carotene. Yellow rice is being used to help eliminate Vitamin A deficient diets in several regions of the world. The gene for beta carotene can be easily added to the rice varieties that will be used in ALS systems. A similar approach can ensure that other essential micronutrients will be present in ALS crops.

One of the concerns of some anti-GM activists is that these designer plants will spread their "unusual" genes to native plants and there will be no way to control the spread of this human designed, genetic information. On the moon or Mars, there are no native plants so this concern isn't an issue. Further, plant production units will be isolated from each other because different types of crop plants require different environments to optimize their growth and food production. Multiple plant production units would also reduce the potential for a serious breakdown that might affect the entire ALS program. It would be easily possible to utilize different genetic variables in separate ALS units. We will need plants that are designed to grow well in decreased pressure environments, with a different gas mixture than we have here on Earth. These plant varieties can be produced on Earth and their unique seeds easily shipped to the colonies.

What we will need to acquire is the ability to grow plants that provide the best nutrition possible for the explorers and the animals that accompany them. Plants will need to have a low percentage of non-digestible waste. It is likely that GM plants, and perhaps animals, will be a mainstay of off-Earth agricultural food production units. Criteria established for plants include rapid growth, diverse and easily harvested foods with a low percentage of non-usable or inedible biomass. For the most part, inedible biomass is composed of cellulose and lignin not directly used by humans.

Some plants that are being considered for food crops, in addition to the salad materials mentioned earlier, include dwarf rice and dwarf wheat, both Irish and sweet potatoes, and various varieties of soybeans. These plants represent the core, but they are just a start. As colonies mature, there will be an interest in making other vegetables and fruits available. Early habitats will not have the luxury of growing plants in large areas. Stacked layers of dwarf plants will maximize space utilization. Several species of dwarf wheat have been developed that can be grown in a much smaller space than "regular" field wheat. One problem with many dwarf varieties, at this time, is that their yield is decreased as well as their height (Figures 11.1 and 11.2).

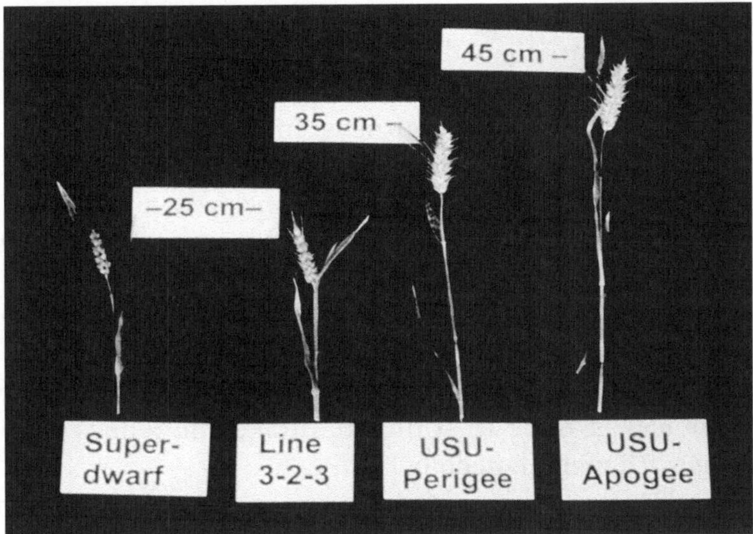

FIGURE 11.1 Some available dwarf wheat varieties. They are much shorter than standard wheat and can be grown more intensively with rows of plant cultivators stacked vertically to maximize the use of available space. One problem with many dwarf varieties, at this time, is that their yield is decreased as well as their height. (Photo courtesy of Bruce Bugby)

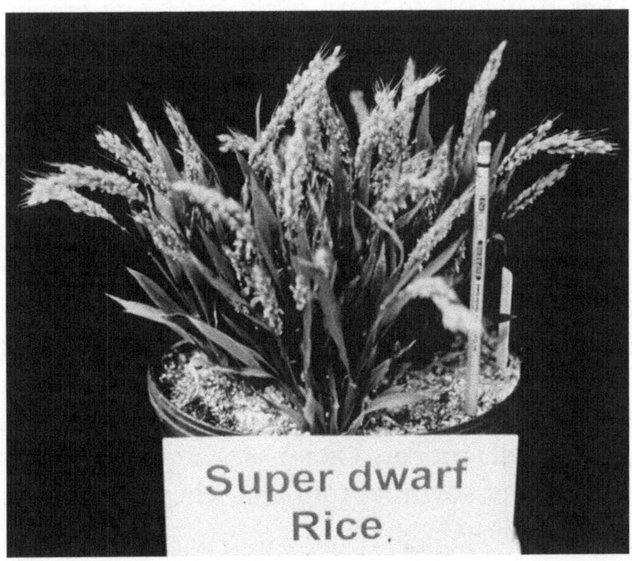

FIGURE 11.2 Super dwarf rice, developed in Japan. Although dwarf varieties can be grown in a smaller space, so far they produce a decrease in yield. The hope is that before they are truly needed to support colonization on other planets, innovative plant growth facilities and genetic manipulation can ensure plentiful and dependable harvests. (Photo courtesy of B. Bugby)

Growing plants that are edible is just the first step in producing food and a nutritious diet. We will need a whole ecosystem with a variety of plants that can be used by us, and the animals that we bring along on our journeys. In developed countries, we have been spoiled by the abundance and selections of plant and animal food available in supermarkets and restaurants. This will change on the new frontier, as there will be much less variety in the raw products available. Colonists will have to rely on ingenuity and innovation to turn a relatively few of these freshly harvested plants as well as animals into not just edible, but tasty meals that will satisfy a diverse audience. It will be important to provide different choices in order to stimulate intake and relieve potential problems with menu boredom.

The ability to grow and to harvest a selection of plant crops is only the beginning of the job of supplying voyagers to other worlds with a diverse, nutritionally adequate, and esthetically acceptable selection of foods. The next step, and it is a big one, is to convert the few basic foodstuffs that will be available, into a variety of delicious dietary products. We have grown too used to buying prepared, ready to eat or cook items at the store. The general public no longer has the skills to work with many raw products. Small scale food processing expertise will be a necessary skill in new off Earth settlements. The greatest potential problem will be with short term visitors used to the variety of Earth foods that are available 'back home.' Old hands and second generation settlers will view the more limited dietary choices as normal.

One way to use the grain crops, wheat and rice, and soybeans will be to turn them into flour using a compact mill. Another option for processing soybeans would utilize equipment whose prototype has been developed at Johnson Space Center. The machine is called STOW. This acronym stands for soymilk, tofu, okara, and whey. All can be produced from soybeans with minimal equipment adjustment. Okara is a non-soluble component left after removal of soymilk, tofu and whey. It can be converted into bread or cookies.

Extruders are used commercially on Earth to turn flour products into the many kinds of pastas available in grocery stores. Small extruders could be used to do the same thing with wheat and rice flour in the food processing section of our new settlements.

A variety of pastas could be made for use in preparing different meals, using imported flavors and spices to provide variety. These flavors could be part of the supply/re-supply program from Earth. As with seeds, a collection of many different spices would only weigh a few pounds, yet greatly improve taste variety in meals. The longer colonies exist, the more sophisticated the agriculture system and food production units will likely become. A continual goal will be not just nutritional adequacy, but also variety and consumer acceptance. One dietary component that will initially be in short supply will be fresh fruit. Strawberries will likely be among the first fruits to be produced as their cultivation requires minimal space or volume and a short interval from planting to harvest.

Many food crops will be grown by hydroponics or in an artificial medium that provides plant nutrients. One product of this kind is called zeolite. Zeolite is hydrated aluminum silicate with no specific texture or shape. It is commonly used as a growth medium for plants. The nutritional content of crops grown, using these methods, can be varied depending on the composition of the plant nutrient solutions used in the hydroponic system and the genetic modification of the plant.

In order for plants to form seeds to produce the next generation and to provide nutrients for humans and other animals, they will need to be pollinated or fertilized. Fortunately, most of the crops that are planned to be used in an ALS system can either self fertilize or are fertilized by the movement of air. These crops include cereal grains such as wheat and rice and also beans, peas, and lettuce. By vibrating plants or using airflow to move pollen from one flower to the next, fertilization can occur in many plants that might otherwise require insect pollination.

As agricultural production becomes more established in off-Earth locations, it would be a great idea to import honey bees to act as pollinators for those plants that do not effectively self-pollinate. As a long time raiser of bees, their product is a great addition to almost all diets. Plus on both the moon and Mars you wouldn't have trouble with bears consuming the harvest as has happened to me. If there were periods of time that blossoms and nectar were not available, the bees could be fed glucose from hydrolyzed cellulose. Under such circumstances, they would need a protein supplement to replace the pollen.

Some plant products, like cellulose, are better used to feed animals other than humans whose digestive tracts can break down cellulose. These animals can then be used as a protein source for the colonists. Animal protein production will begin early in colonization. It will require an existing plant growth program that can provide plant waste products. Some of the plant wastes including cellulose may be "fed" to simple organisms like bacteria and algae. These organisms, in turn, can be processed into animal food. The first animals to be raised in developing outposts and settlements will likely be fish and fowl.

Tilapia is the common name for a group of fish native to Africa that are now widely grown on fish farms all over the world. They grow well in these aquatic farms and can help dispose of plant wastes not used by humans. Chickens have not only the capacity to provide eggs with a high nutritional value but are extremely efficient in converting their diet into food usable by humans: pound of meat produced per pound of feed consumed. This well-established high rate of dietary conversion into animal protein is the reason that chickens and turkeys are the least expensive meat in our grocery stores. At this time, there are problems with avian species in μg, as noted earlier with pigeons, chicken eggs, and quail hatchlings. However, there is no reason to suppose that birds will not be able to successfully adapt, grow, and reproduce in reduced gravity.

Planning for the production of chickens and fish in a decreased gravity field raises a question. How much muscular development will occur in decreased gravity? Based on our knowledge of biological adaptations, here and in space, it will probably be a quite different response for these two kinds of animals. Both grow rapidly and provide valuable animal protein for human diets on Earth, but their response to a changed gravity field will be different. Let's start with chickens.

Raising chickens for food is based on the assumption that they will convert their diet to muscle protein. Birds do require an antigravity skeleton and supporting muscles that allow them to function on Earth. However, their skeletons are less dense than ground dwelling mammals as they balance light bones to allow flight with enough structural strength to prevent fractures when walking, running, or landing. It is probable that leg and thigh muscles will be decreased in chickens in a lower gravity field, but very little change in breast and wing muscles unless the chickens start flying more because of the decrease in body weight. This assumes that habitats are large enough to permit flying. Just imagine a 3 pound chicken transported to the moon. It would suddenly weigh only one half pound. Those largely non-functional Earth wings could easily support flying on the moon inside the habitat or colony. Chickens could not exist outside any easier than humans, and a chicken space suit sounds like a joke. Regardless of that, an order of "wings" on the moon will be no different than one here on Earth. Their leg bones will also diminish as body weight is decreased. Based on many hyper-gravity centrifuge experiments with chickens, they should be somewhat taller with spindly legs. There is no reason to suppose that birds cannot reproduce and develop in a decreased gravity environment and provide an important part of our diet on the moon or Mars.

On the moon or Mars, fish will continue to have to swim in a dense water environment. They may have a minor change in their swim bladders to ensure that they are neutrally buoyant but slightly denser than their water environment so that they slowly sink. For a fish to swim should be no different from a muscle or energy expenditure standpoint, than swimming on Earth. Fish do not use their bones or muscles as a support to defy gravity on Earth and will not need to do so elsewhere. Without antigravity muscles, their body composition should not change. They will continue to swim in a watery environment in the same way no matter where they are. Undoubtedly they will continue to require some gravitation or a light field, in order to establish up and down, like the fish on Skylab. In transit to the moon or Mars, an enclosed aquarium may be needed to keep the fish and water contained.

Development of an operational ALS food production system on the moon or Mars must be preceded by building successful production units on Earth. If we can't accomplish plant growth, animal growth, food processing, and waste recycling in a semi-closed or closed ALS system on Earth, we will not be successful in an outpost elsewhere. It will take time and require carefully controlled and monitored biological production systems to accomplish this feat on new worlds. As we begin to move in this direction the bottom line to be remembered is that if it doesn't work here, it won't work away from Earth.

On site food production for our space voyagers will be a requirement, but we need to develop the skills to make it happen. Even with experience in closed systems on Earth, the new breed of explorers of our neighboring worlds will need to be innovators, eager to invent and develop unique approaches to the age old necessity of feeding the populace.

12. Speculations

We are the offspring of history, and must establish our own paths in this most diverse and interesting of conceivable universes – one indifferent to our suffering, and therefore offering us maximal freedom to thrive, or to fail, in our own chosen way.

Stephen Jay Gould (1941–2002), Paleontologist, Evolutionary Biologist, Philosopher, *Wonderful Life*

Living in Space? Living on the moon? Living on Mars? The real question is not if, but when. Our race has a deep and abiding urge to move outward: to explore, to solve mysteries of the unknown, to conquer new worlds. We have pretty well overrun this planet. Our current and approaching technologies have and will provide the long awaited opportunity for Earth life to expand outward to new worlds. We are not destined to remain earthbound. We will explore and settle both the moon and Mars. Our grandchildren, their children, and especially their grandchildren will be our new pioneers and lead us to new destinations.

There is one thing to be wary of in making plans to expand human life to new worlds. It is '*traditionitis.*' One hundred years or so ago, those infected by the to 'traditionitis' bug did not want cars on the road because they would scare the horses. In ancient times the stay-at-home advocates would wave goodbye to those who left Africa to begin to explore this world. New reasons will undoubtedly be brought to the fore to block our destiny of expanding to other worlds. "We've never done that before!" The appropriate response is to pat them on the head and continue to expand our presence to new parts of our Solar System.

We and most, if not all, of the plant and animal life on Earth will be able to adapt to not just the lower gravity of the moon and Mars but in most cases to μg as well. The success in selecting non-looping Medaka fish to live in μg and the relative speed that small fish adapted during Skylab are good omens for our future attempts. Birds, as yet, have not really been given an opportunity to adapt.

R.W. Phillips, *Grappling with Gravity: How Will Life Adapt to Living in Space?*, Astronomers' Universe, DOI 10.1007/978-1-4419-6899-9_12,

That must wait for longer-term avian space journeys. Based on all other vertebrate species with some restraint in the first days off exposure to μg they will adapt.

One facet of the unknown deals with using artificial centrifugal gravity in low gravity environments. If the desire is to be able to return to Earth's 1*g*, it will be necessary to curtail the rate of adaptation to decreased gravity. Using the moon as an example, what factors need to be considered? The desire may be to readily move from moon to Earth and back instead of having prolonged periods in the lower gravity field. In such a case, it will be desirable to attempt to block or greatly decrease the rate of adaptation.

Using a centrifuge on the lunar surface to decrease adaptation would present many problems. Perhaps a 1*g* or even 1.5–2*g* centrifuge could be used for periods of time each day, but this would not stop lunar adaptation during the rest of the day. Until permanent lunar living becomes a reality, intervals of return to Earth may be necessary. The μg environment of space or the low gravity environment of the moon or Mars will be insidiously constant in causing adaptation and must be continuously circumvented to be successful. Programs designed to establish permanent settlements away from Earth must be mindful of the fact that Earth life is very adaptable. It does not appear that you can be simultaneously adapted to more than one gravitational field at a time.

As an alternative to returning to Earth, some individuals may opt to make the moon a permanent residence. They will raise families and make the moon their home. They will not need to develop elaborate systems to circumvent adaptation to low gravity. Their offspring may look a bit different to Earthlings and they may well find it difficult to visit Earth from a physical perspective. Such individuals will be blazing a trail for the expansion of humankind beyond our home planet. The same scenarios would need to be developed for a life on Mars after experience has been gained on the moon. With greater mass and increased gravitational field, the measures needed to simulate Earth gravity on Mars would be less rigorous. It will be less complicated to maintain reasonable Earth fitness on Mars than on the moon.

When a return to the moon is mentioned, the question is often raised, "If it is so important to go back to the moon, why haven't we done so already?" After all, the Apollo program successes and moon trips were 40 years ago. Historically, there is a basis for this seemingly slow rate of progress. Major explorations are usually initiated by governments and then taken over by the citizenry. Settling in a new place to live is quite a different ball game from exploration. It takes time. Think about the settling of North America by Europeans. Columbus sailed in 1492. The first permanent settlement established by Europeans in what is now the United States was Saint Augustine, Florida, in 1565, 73 years after Columbus's first voyage. The first northern European settlement was Jamestown in 1607. This is 115 years after Columbus. The pilgrims did not land at Plymouth Rock until 1620, 128 years following Columbus. The intervening years were spent in exploration and discovery. With these thoughts in mind, we can imagine that a settlement on the moon is likely in decades not years. Be patient, it is going to happen.

Exploring and settling new worlds has never been easy. The first northern European attempt at settling North America was at Roanoke Island on the North Carolina outer banks in 1587. Three years later there was no trace of the settlers. The end of the "Lost Colony" is still a mystery, although there are many suppositions and theories. In 1607 Jamestown, the first successful permanent English colony in North America, was founded. By 1624, a total of 7,500 settlers had arrived, but only 1,100 were alive in 1625. Six out of seven perished. Disease, wars with Indians, and starvation all took their toll. The pioneers that founded the Plymouth colony at Plymouth Rock, Massachusetts, left England with 101 pilgrims. They arrived in early December in the year 1620, an inappropriate time to begin farming in a new world. Approximately half died that first winter. Planning to start a new life in a strange and unfamiliar world is not for the faint hearted.

Pioneers starting colonies on the moon and Mars will not have to deal with hostile inhabitants and unanticipated disease, but food will still be a problem. Until a functional agricultural system is in place, plentiful food may be a problem, particularly on Mars where re-supply will be so much more difficult.

Captains Lewis and Clark led a military contingent from St. Louis, a frontier settlement, to the Pacific and back in 1804–1806. They mainly explored portions of the Louisiana Purchase and the far west, via the Missouri river and then the Columbia. In the areas

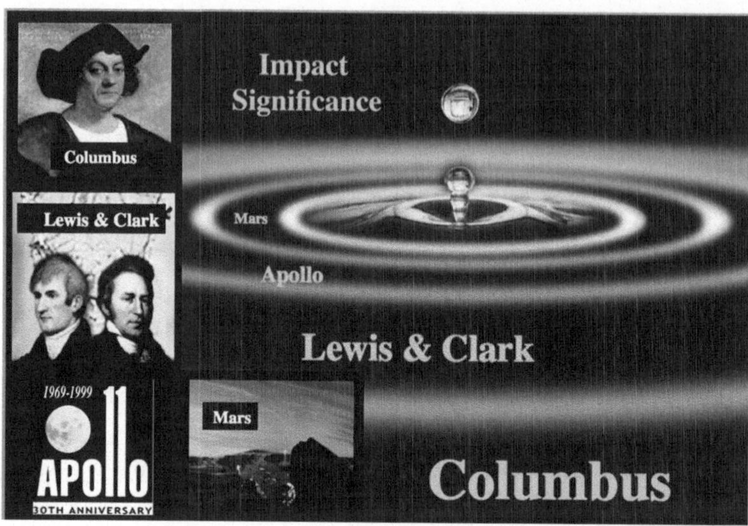

FIGURE 12.1 The impact and significance due to Columbus' voyages has now expanded for over 500 years transforming North and South America. The result of Lewis and Clark's expedition 200 years ago has been to transform the American West. We are 40 years past the Apollo voyages and ready to go back to the moon to stay. Time will tell how long it will be before humans first visit and then begin to reside on Mars.

that they traversed, the first to become a state was Missouri in 1821. From a standpoint of miles traveled, the greatest component of their trip was a portion of the USA's Northwest. That area became the states of Washington, Montana, North Dakota and South Dakota in 1889, 85 years after Lewis and Clark first passed through. In comparison, the Lewis and Clark expedition was the exploration of a portion of a single continent, not a whole new world. Figure 12.1 illustrates that the impact of momentous events is not rapidly felt. The impact and significance of the Apollo missions to the moon has yet to be fully appreciated, while human exploration of Mars has yet to begin.

The important thing is the long-term picture. Figure 12.2 is a depiction of cause and effect. Ripples from that first Columbus voyage have spread over two continents in the last 500 years. In the 200 years since Lewis and Clark began the opening of the west, the United States has moved from sea to sea. In the almost 40 years since humans first began exploring the moon, unmanned probes from several countries have added to our knowledge

a

FIGURE 12.2 (**a**, **b**) This ambitious space elevator project, when completed, could make rocket launches a thing of the past. Using a nanotube ribbon and laser power, a tractor can lift cargo and passengers into space at a few hundred mph. After arriving at the space terminal, one would transfer to another vehicle for travel to the final destination. (Photos courtesy of Brad Edwards)

b

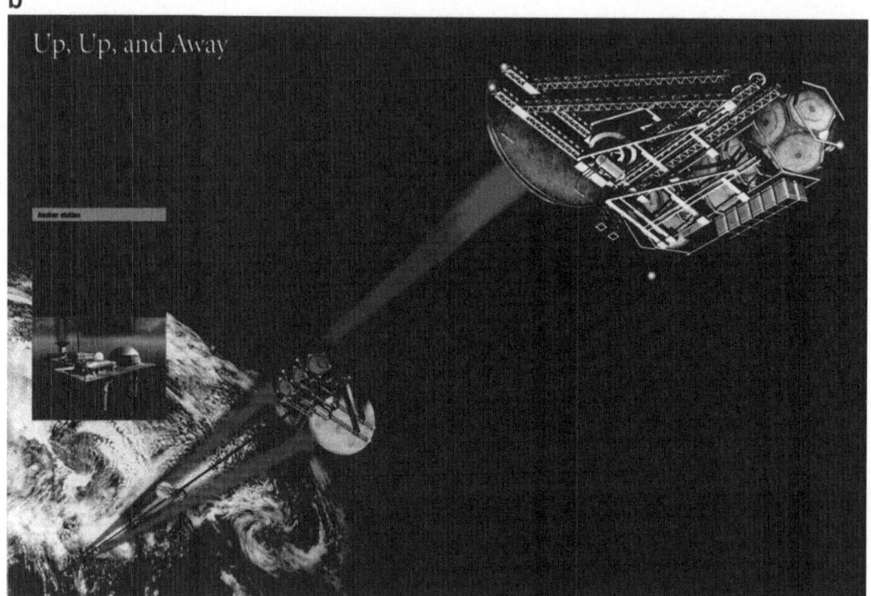

FIGURE 12.2 (continued)

of the moon and Mars. They are preparing a path for human exploration and discovery that will pave the way for future colonization. Historical perspective is an integral component of predicting the future.

An important consideration that must be kept in mind is that our history of exploration and expansion really rests on individual endeavor not governmental fiat. Successful settlement of other worlds will likely be by individuals even though they are initiated with governmental support. It is not the explorers always eager to push the horizon back, but the settlers determined to make a new home that will populate the first colonies on the moon and Mars. As the colonies reach a stage of advanced development, they will desire to become independent, spread their wings, and form a new society appropriate for their particular and unique situation. The job of Earth's majority should be to encourage and support their efforts, not dominate them.

In the near term, we will continue to experience μg in LEO. Later, we will leave Earth's vicinity in transit, first to the moon and back, with months of exposure to a lesser gravity. Later, travelers will face longer term voyages to Mars, 6 months for a one way trip. These two celestial bodies are the next logical step. They will become our second homes and will precede some of the grander plans to build cities in the sky or establish settlements in other regions of the Solar System. The cost and technological innovation necessary to construct a large torus, in essence an entire floating city at a Lagrange point, must wait till we gain more experience in living away from Earth. A torus or other human-built structure in space will have no natural resources. Everything must be imported and continue to be transferred to such a sky world as its atmosphere slowly leaks into space.

The more reasonable approach is to capitalize on our forthcoming experience on the moon, to make it our first permanent extraterrestrial colony. With a combination of mining for helium 3, tourism and senior living quarters, visual and radio astronomy, and ease of launching payloads into Earth orbit and beyond, there is much to be gained by expanding our activities and technical achievements to this close neighbor. Individuals may choose to be transients with perhaps 1–2 year tours of duty or opt to become permanent residents. Their first allegiance will be to ensure the safety and well-being of other citizens of the moon.

Perhaps the proposed 'space elevator' will become a functional reality. If it is successfully developed, leaving Earth for the cosmos will become much less expensive and more available to many. It would greatly facilitate the transfer of supplies and equipment to both the moon and Mars. The concept of a space elevator is based on ultra strong nanotubes made from carbon. This material is so strong that only a thread is needed to support an automobile. The concept is simple. Imagine a small rope with a weight on the end. If you were to swing the rope and weight in a circle over your head, the weight would never respond to Earth's gravity and fall to the ground. It would be held in place by centrifugal force. The space elevator is designed on the same principle. The base site would be anchored on the equator in the Pacific Ocean. A thin strand of nanotubes would connect the base with a space platform in geosynchronous orbit at the upper end. Then it will be added to, eventually making a ribbon about 3 ft wide and 60,000 miles long. The elevator or climber would be sent up the ribbon carrying large loads using laser beams (Figure 12.2). Many believe that the elevator is technically feasible and will be a major means of future space access.

Moving away from our home planet to permanently reside in lesser gravity fields or live in a rotating habitat with its own unique gravity is technically and biologically feasible. Early voyagers may desire to return to Earth and must, therefore, be prepared for prolonged re-acclimation, particularly after long stays on the moon. Returning to Earth's gravity from Mars will be less troublesome, but the length of the trip will be more tedious.

Time will tell how long it will take us to establish colonies with plants and animals that are well adapted to life in lower gravity fields. Our knowledge is such that we can predict that with few constraints, advanced life is capable of spreading outward from this planet. An Earth-like atmosphere from the standpoint of nitrogen, oxygen, carbon dioxide, and water will be needed inside habitats and colonies. Atmospheric pressure and nitrogen content can vary depending on location and experience. Moderate temperatures must be provided in living and working quarters. Based on our experience in microgravity, life can adapt and prosper in the gravitational fields on the moon and Mars. Certainly life has evolved and prospered over millions of years in both relatively mild and severe climates that are here on our home planet.

It is easy to lose track of our relevance if you consider the grand scene provided by the universe. Earth is a minor planet in one non-remarkable Solar System in an arm of our galaxy, one of billions of solar systems in countless galaxies. We are the temporarily dominant species on our world and have been for a few thousand years out of the millions of years that life has existed on planet Earth. We sometimes have difficulty placing our race in its appropriate perspective.

From an entirely different perspective, it can be difficult to differentiate the very large from the very small. Below in Figure 12.3 there are two images. One is microscopic and the other macroscopic, taken by the Hubble Space Telescope. Can you tell which is which? Cover the legend before examining the two pictures.

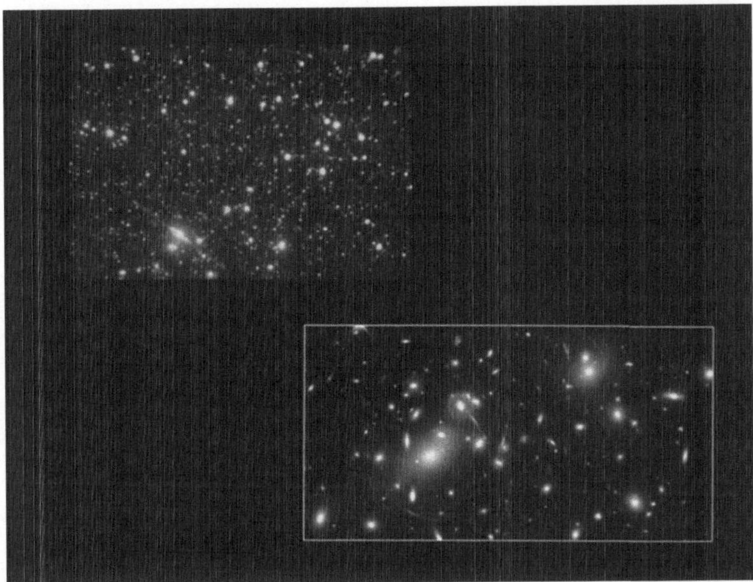

FIGURE 12.3 *Upper left,* is a microscopic view of a drop of seawater. Small "stars" are marine viruses, larger structures bacteria, the long "galaxy" is a diatom. Diatoms are single cell, usually photosynthetic algae. The *lower right* picture is a deep space image from the Hubble Space Telescope. (Images courtesy of Millie Hughes-Fulford)

Earth is indeed a paradise compared to the worlds that we are planning to settle, but the human spirit embraces the mystery of the unknown, of exploration, and of discovery. We will reach out to our neighboring worlds and find once again that expansion of our species into new domains is worth the effort and cost. Even a lower pressure artificial atmosphere that will be standard inside an outpost or colony would become familiar after adaptation.

In the foreseeable future, explorers or residents on Mars, with a small telescope, will be able to see a picture similar to the one below taken by the Mars Global Surveyor in orbit around that planet (Figure 12.4). Our blue Earth, with its circling moon millions of miles away, is visible and beckoning. Are you ready to join past explorers and pioneers as we expand to new worlds?

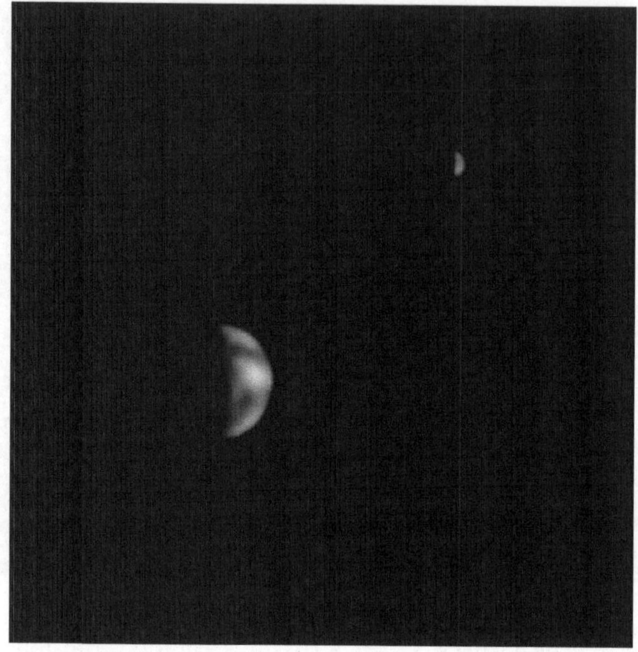

FIGURE 12.4 A picture of the Earth and its moon as seen by the Mars Global Surveyor, May 2003, in orbit around Mars. (Photo courtesy of NASA)

ERRATUM TO:

Taking Your Body to Space

Robert W. Phillips

R.W. Phillips, *Grappling with Gravity: How Will Life Adapt to Living in Space?*, Astronomers' Universe, DOI 10.1007/978-1-4419-6899-9,

DOI 10.1007/978-1-4419-6899-9_13

In chapter 7 ("Taking Your Body to Space"), page 125, the presentation of figure 7.17 is incorrect. The correct figure is given below.

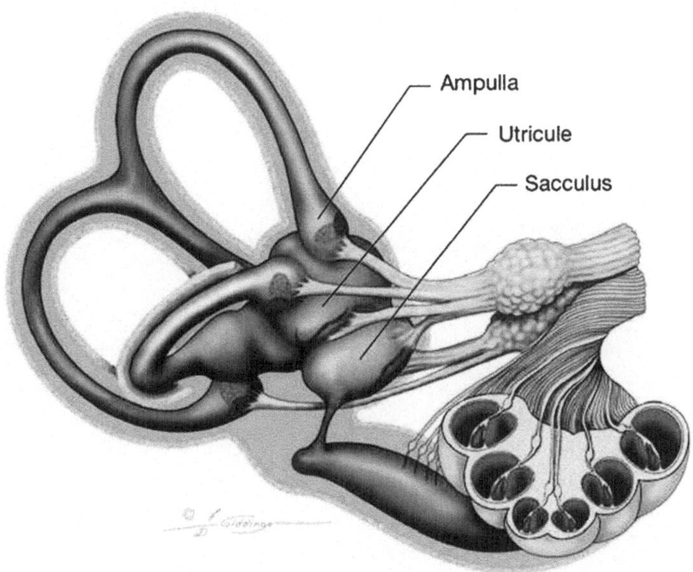

FIGURE 7.17 Vestibular apparatus. The sensory receptors are in the ampulla at the ends of the semicircular canals. The utricle and saccule detect position with respect to gravity. The yellow nerve carries sensory information to the brain.

The original online version for this chapter can be found at http://dx.doi.org/10.1007/978-1-4419-6899-9_7

Appendix A

Description of Videos For

Grappling with Gravity: Adaptation Is the Key to Success

The videos are listed as movies in quick-time and are arranged in numerical order that matches their location in the book. A quick-time installer is also present with the video segments.

Video #001 Skylab Exercise page 113

Exercise on Skylab in the 1970s was visually more interesting than using bicycles, treadmills, and resistance devices as is done on the Space Shuttle or ISS. On Skylab with its greatly increased available space, the crews had a unique exercise opportunity. They could easily combine acrobatics with running as is demonstrated in the video. The ability to vary exercise modalities would make it much less onerous.

Video #002 Rat Behavior page 120

The video is in four sections. The first section is a rat that has been removed from its cage after a number of days in spaceflight. He moves easily and confidently around and upon the astronaut's hands as well as on the surface of the cage. When briefly free floating, he exhibits the typical behavior of rats when exposed to μg by extending all four limbs looking for a solid surface to grasp. At other times, he appears completely normal.

Section 2 is a picture of a normal ground control rat that has been placed in a 10 gal aquarium for observation. He walks confidently around and examines as much of his location as he can by standing up in corners of the aquarium.

R.W. Phillips, *Grappling with Gravity: How Will Life Adapt to Living in Space?*, Astronomers' Universe, DOI 10.1007/978-1-4419-6899-9,

Section 3 shows a rat that had been in space for 9 days. He is either the one shown in section 1 or another of the same group. He was placed into the fish tank just 2 h following landing and re-exposure to Earth gravity. It is hard to imagine how he could have been less mobile. He does not use his limbs to support his body off the bottom of the tank. What little mobility he exhibits is primarily by use of his forelimbs not his hind limbs. He made one very feeble attempt to investigate in a corner of the aquarium but was not able to stand on his hind legs. This drastic change in muscle ability resulted from only 9 days of adaptation to μg.

Section 4 is a rat from the same space mission that has been back on Earth for 1 week. He has recovered a lot of his mobility and is again able to stand and inspect the corners. His gait is not completely normal as he carries his hind limbs more forward under his body. It is likely that given more time he would have completely recovered.

Video #003 Rotating Circle page 132

We are so ingrained with the knowledge that light comes from above that simple shading of a circle can establish the belief that we see concave or convex. Watch the change between concave and convex as the circle rotates. This is relevant because on the International Space Station, as an example, light only comes from the two corners that are the nominal up. Yet the ISS has equipment and laboratory facilities on all four sides. This type of new environment can create illusions that must be corrected by adaptation.

Video #004 Jellyfish page 144

Jellyfish normally float upright with the gravity receptors on the bottom. They pulse their mantles as needed to maintain the proper depth in the ocean. When exposed to μg they were either quiet or pulsed in large loops. In Earth's oceans they would never pulse downward or in loops. After the end of the mission, their behavior was normal.

Video #005 Medaka Looping page 145

When Medaka are first exposed to μg on parabolic airplane flights, they rapidly turn tight somersaults, spirals, and loops.

Although not evident from the movie, a few of the fish behaved in a normal way during the μg period. They were separated from the other fish and a strain of non-looping Medaka was bred. This behavioral characteristic bred true and the new stock was the basis for selection of four non-looping fish for spaceflight.

Video #006 Skylab Fish page 146

Small killifish called Mummichogs were carried to Skylab. When they first arrived, they swam in somersaults, loops, and spirals in the same manner as Medaka. Eighteen days later, 19 after entering μg, they had ceased to swim erratically. They had adapted to space flight by learning to disregard vestibular function or perhaps dysfunction. They relied on light as an indication of up and down. Remember, the plastic bag aquarium was velcroed to a vertical wall in the lab. They swam with their backs to the laboratory where it was light and their abdomens towards the vertical wall. Their speed of adaptation is a good indication that fish will be able to adapt not only to decreased gravity on other planets but also to μg for perhaps a 6 month trip to Mars.

Video #007 Tadpoles page 147

This video is of a group of tadpoles on the KC-135 airplane during a parabolic loop creating a period of μg. Prior to entering μg the tadpoles were all on the bottom with their heads pointing down. During μg they left the bottom and began to swim erratically in loops and somersaults. This continued until the period of μg was over and they returned to the bottom of the container.

Video #008 Pigeon page 148

A pigeon was released during the μg portion of a parabola on the KC-135 airplane. It flapped its wings and rotated in place without any forward motion. The bird did not seem to have any ability to control its behavior. No observations were made of the animal while standing in its cage during a parabolic loop. Based on this example birds will not be able to initially function without restraint in a μg environment. There is no basis for determining how long it may take to become adapted to spaceflight.

Video #009 Chick page 148

Quail chicks were hatched on the Russian Space Station Mir. One of the birds is shown in this sequence. It was removed from the incubator for observation. If held in a cosmonaut's hand, it would peck particles of food from his hand and wrist. It did not appear to be bothered by μg as long as it was restrained. When it was turned loose, it rapidly rotated, turning in somersaults, loops and spirals at about one revolution per second. The chick continued this behavior until it was physically stopped by catching it during its rotation.

Video #010 Medaka Breeding page 163

Four Medaka from the non-looping strain selected in Japan were chosen as the flight animals. They were all flown together in a small aquarium on the Space Shuttle. Medaka have an uncommon method of breeding for fish. The male attaches himself to the female using papillary processes on his anal fin. By rubbing her abdomen with his body, he stimulates her to release eggs in a gelatinous mass on to her abdomen. He ejaculates sperm and the eggs are fertilized. They are released from the female's abdomen, progress through embryogenesis and hatch.

Section 1 of the video shows two Medaka breeding on Earth. There is a great deal of activity by their fins during the process.

The next section, again on Earth, shows embryogenesis. Beating of the heart is very obvious. The two large dark eyes in the embryonic fish can be clearly seen. As the embryos mature, they break out of the egg and swim away.

Section 3 shows the four fish selected for spaceflight during the mission. It is easy to tell that it is in space as the large air bubble does not float to the surface. One male and one female are breeding on the right of the bubble. The second male is anxious to participate but the female is not receptive. The sequence continues with a number of fertilized eggs that are obvious with the prominent dark eyes. The last picture in this section is of a single newly hatched Medaka. Actually, a number of Medaka were hatched and returned to Earth where they matured and produced additional generations of 'space fish'. They were all apparently normal.

The last section of this video shows a male and female breeding to the left of the air bubble. The second male is disturbed by not participating. He could be considered to be in a jealous rage. By butting the reproducing pair with his head he is eventually able to break them apart.

Video #011 Quail Incubator page 169

This video shows the group of Japanese quail that were hatched on the Mir Space Station. As a group they were continuously flapping their tiny wings and rotating in an uncontrollable fashion. They were unable to stop gyrating and hold on to the wire mesh in the bottom of the incubator. This is not completely surprising as new born quail, although precocious, normally walk flat footed on the ground. They do not perch nor roost in trees. Quail cannot fly for some time after birth. Immediate exposure to a μg environment was too much for them. They did not have the opportunity to adapt. A restraint device to immobilize newly hatched chicks has been developed, but has yet to be flown in space.

Video #012 Rat Pups STS 72 page 173

The first segment is of rat pups that were 5 days old when launched into space. Few of them survived. It was difficult for the mother to control and nurse her aimlessly floating offspring. On several occasions she reaches out and grasps a young pup and pulls it back toward her mammary glands. Most of the pups that were 8 days old when launched survived even though their eyes and ears were not open. Their motor skills had matured enough that they could nurse the mother when they came in contact with each other. The last segment shows rat pups that were 15 days old at launch. Their eyes are open and they can locate their mother and move to her. All 15 day old pups survived. As you watch them move about, their motor control is much greater and they have a well developed hair coat.

Video #013 Apollo page 189

There are four segments to this video, each showing an Apollo astronaut working or exploring on the surface of the moon. As astronauts move about the surface, their gait is between a hop and a shuffle. Their space suits weighed 180 pounds on Earth and would weigh 30 pounds on the moon, so that the astronaut and his suit would weigh 55–60 pounds. Because of that decrease in their accustomed weight, they could perform movements not possible on Earth in spite of the awkward and cumbersome space suit.

Segment 1, **Stumbles**. The astronaut is taking a regolith sample from several feet under the surface. He inserts the sampling rod all the way, indicating that there were no rocks immediately under the chosen spot. As he completes the insertion, he stumbles and falls forward, catching himself with his arms. He extends his arms, bounces back to his knees and again becomes erect. He then removes the sampling rod on the second try. In several of the segments, it is obvious that it is not easy to grasp an object at or near the lunar surface.

Segment 2, **Hammer slips**. In this segment an astronaut is again attempting to acquire a lunar sample. In this case he uses a longer rod that needs to be pounded into the ground. He places the rod in position and begins to pound with a hammer. The rod seems to have hit an obstruction as the hammer no longer has an obvious effect. Then the hammer slips out of his grasp and bounces across the lunar surface. He makes four attempts to retrieve the hammer jumping higher off the ground in each try and then bending his knees upon landing to get closer to the hammer. He never succeeds by the end of the segment.

Segment 3, **Dropped bag**. The astronaut is coming out of a crater or a valley climbing a hill to retrieve a bag he had dropped. He grasps the bag on the first try, but falls to his knees. After getting to his feet, he continues up the hill spinning around near the end of the segment.

Segment 4, **Stumbles several times**. This segment begins as the astronaut stumbles or goes down on his knees to retrieve a bag. He stumbles backward while getting up, then drops the bag and falls forward. He appears to drop the bag again, adds another bag to it, and moves off away from the bags.

These video segments indicate that heavily suited astronauts can move freely around the lunar surface even before they have had the opportunity to adapt to this new environment. The lunar module is much too small to walk in so that the only walking experience that the Apollo astronauts had is the brief time they are out of the lander and moving about the surface.

As we return to the moon in the near future, it will be for longer periods of time, months not days. It is quite possible that the lunar space suits will not be as cumbersome as those of the Apollo era. With a much longer exposure, the lunar crews will begin to adapt to this new environment and acquire the skills needed to become more mobile on the surface. As outposts develop with a permanent human presence, astronauts will also adapt to moving around in an inside environment without cumbersome space suits.

Appendix B

Determining the Strength of the Gravitational Field While in Space

It is easy to determine the strength of the gravitational force at any distance away from Earth using a simple pocket calculator. It gives an approximate value for the gravitational field of a space ship in orbit around the Earth or at any distance. It is the square of the Earth's radius, divided by the square of the distance from the center of the Earth to the satellite, expressed as a percent by multiplying the answer by 100. The Earth's radius is approximately 3,960 miles. If a space vehicle were flying 150 miles above the Earth, the distance from the planet's center to the vehicle would be 3,960 + 150 or 4,110 miles. This number is only slightly larger than the Earth's radius. Using the calculator, square each of the two numbers. Now have your calculator divide the square of 3,960 by the square of 4,110, this provides a number of 0.928 or a percentage of 92.8. So traveling in LEO 150 or so miles above the surface as many human crewed vehicles do, the gravity field is about 90% of its value on Earth. If you made the same calculation for a spaceship 1,000 miles above the Earth, the Earth's gravitational pull would still be almost 64% of its value on the Earth's surface.

Acronyms

ALS	Advanced life support
AU	Astronomic unit
CME	Coronal mass ejection
CNS	Central nervous system, brain and spinal cord
DNA	Deoxyribonucleic acid
ECF	Extracellular fluid
EKG	Electrocardiogram, also (ECG)
EPO	Erythropoietin
ESM	Equivalent system mass
Fe	Iron
GCR	Galactic cosmic rays
GM	Genetically modified
ISS	International Space Station
LBNP	Lower body negative pressure
LDEF	Long duration exposure facility
LEO	Low earth orbit
μg	Microgravity
MPLM	Multi purpose logistics module
NASA	National Aeronautic and Space Association
NOAA	National Oceanic and Atmospheric Association
rbc	Red blood cell or erythrocyte

SAC	Short arm centrifuge
SAS	Space adaptation syndrome
SMS	Space motion sickness
SPS	Space power system
STOW	Equipment to convert soybeans into soymilk, tofu, okara and whey

Glossary

Adaptation When used with regard to living systems, it means a change in structure, function, or behavior to become more fit to prosper in a new environment.

Advanced life support A system designed to create an ecological system patterned after Earth; food will be grown, carbon dioxide utilized and oxygen and clean water produced, also known as ALS.

Amatorial Pertaining to love.

Alpha Centauri Closest star system to our sun, it is really a binary system, consisting of two stars. The Alpha Centauri complex is over 4 light years from here. Using current technology, we cannot travel to or send probes to the vicinity of these stars.

Ampulla A sac at the end of a semicircular canal in the vestibular apparatus in the inner ear. It detects angular acceleration.

Antibody A protein produced by immune cells that can counteract a virus or bacteria.

Antimatter propulsion A functioning antimatter propulsion system has yet to be devised. The principle behind antimatter propulsion is the production of negatively charged protons. When they interact with normal positively charged protons, an immense amount of energy is released.

Apogee The high point of an orbit, the furthest from Earth

Approbation Consent or approval of a plan or action.

Astronomical unit Mean distance from the Earth to the sun, about 93,000,000 miles, used as an acronym, 'AU'.

Bane An action or entity that continually causes problems or misery.

Bends Decompression sickness, refers to moving from a high pressure to a low pressure environment. Bubbles of nitrogen may form in the blood and fluids. These tiny bubbles can be quite painful and dangerous. They can be prevented by breathing a low nitrogen atmosphere.

Bisphosphonates A class of compounds that have been helpful in treating bone loss in elderly patients. Bisphosphonates are being tested in space flight to see if they are effective preventing bone loss.

Brine shrimp Small aquatic crustaceans that can live in highly saline water. Unhatched eggs remain viable for long periods in a desiccated state.

Cardiovascular Refers to the heart and blood vessels as a single system.

Cellulose A carbohydrate composed of glucose or sugar units. It is a major constituent of the cell wall of plants, providing structure in plant growth against gravity. It is not directly digestible by animals, but can be broken down by bacteria in the digestive tract.

Cilium Short hair-like projection from a cell can provide sensory input to the cell and on to the brain. In other cases, movement of the cilia cause cell movement.

Circadian A biological rhythm on a 24 h basis. It may be related to behavior or physiology. Circa means about, dia stands for day.

Colloquially Indicates an informal use of words or thoughts, often used to simulate or mimic conversation.

Convection Movement of a liquid or gas based on differences in temperature. Heat rises creating currents.

Coronal mass ejection An eruption on the surface of the sun that ejects large masses of material, associated with solar flares.

Countermeasure A device or procedure designed to prevent adaptation to spaceflight and keep astronauts in an Earth adapted state.

Deuterium An isotope of hydrogen whose nucleus contains a neutron as well as a proton. It gives up the neutron during fusion when energy is released.

Drosophila A small fruit fly, it is commonly used in biological and genetic research. Fruit flies have been flown in space on several occasions.

Electrochemical The mechanisms used by the nervous system to transmit information, both sensory information to the brain and commands from the brain to body parts such as muscles.

Electron A particle outside the nucleus of an atom that has a negative electric charge.

Electronic nose A device, now being used in space, to detect potentially toxic gases in the atmosphere inside space vehicles.

Endotherm An animal that produces heat that warms its body. Warm blooded, mammals and birds are endotherms.

Ephyrae Free swimming juvenile stage of a jellyfish.

Equivalent system mass A method of using quantitative analysis to determine at what point it is more logical and cost effective to grow foods rather than import them to other worlds. It includes all aspects of decision making, also known as ESM.

Erythrocytes Red blood cells, rbc. They carry oxygen to tissues of the body from the lungs and carbon dioxide to the lungs for transfer to the atmosphere.

Erythropoietin A hormone produced by the kidneys that stimulates the production of red blood cells (rbc) in the bone marrow, often designated as EPO.

Escape velocity Speed required to overcome a planet's or moon's gravitational pull.

Eukaryote A cell whose nucleus and other structures are membrane bound, typical of all plants and animals, but not bacteria.

Extolling Praising, usually with enthusiasm.

Extracellular fluid That fluid within the body that is outside of cells. It includes fluid surrounding cells and the liquid portion of the blood.

Femur The thigh bone, which has a ball on the upper end, that fits into a socket on the pelvis.

Fission Asexual reproduction by splitting into two equal parts. Both develop into new individuals.

Foibles Small weaknesses.

Follicle A small cavity or gland.

Fusion Interaction between two atomic nuclei that releases a large amount of energy. Controlled fusion reactions can generate electricity.

Galactic cosmic rays High energy radiation from throughout the galaxy. They represent a danger to space travelers away from our magnetosphere.

Galaxy A large star system held together by mutual gravitational fields and isolated by distance from similar star systems. Our solar system is part of the Milky Way Galaxy.

Gene A deoxyribonucleic acid (DNA) unit on a chromosome capable of transmitting information to the next generation.

Genetic modification Technology capable of modifying plants and animals genetically so that they have new characteristics not present in their ancestors. It is also called transgenic as it may involve transfer of DNA fragments or genes from one species to another.

Genome The total genetic information passed to subsequent generations. It is the DNA makeup of an individual derived from its parents.

Geothermal Geo represents earth, thermal represents heat. It is the process of utilizing heat from the molten center of any planet. Hot springs are geothermal.

Gravitropism Growth response of a plant to gravity. Shoots exhibit negative gravitropism by growing up. Roots grow down in a positive gravitrophic manner.

Gravity The force of attraction between two bodies causing each to be pulled towards the center of mass of the other. It usually refers to the attraction of large masses like a planet or a moon.

Helium 3 A form of helium formed in and released by the sun. A large quantity of helium 3 has been deposited on the moon. It can be used in fusion reactions to release energy.

Hermaphrodite A plant or animal having both male and female reproductive organs.

Homeostasis Process by which animals maintain a stable internal environment, homeo = similar, stasis = equilibrium.

Hydrostatic Pressure caused by a column of water.

Hypothalamus A structure in the brain that controls basic body functions such as temperature, hunger, and thirst.

Impetus Energy or motivation to accomplish something.

Inert Not able to move.

Inexorable Unable to stop or alter. An avalanche or the tide in the oceans could be called inexorable.

Instar The period between molts in animals that have an exoskeleton like shrimp or insects.

Ion An atom that has an electric charge due to losing or gaining electrons.

Ionize The process of gaining or losing electrons that creates an electric charge.

Isotope A chemical element that has two or more forms due to a change in the number of neutrons. The number of protons and atomic number do not change. In this book the two isotopes are deuterium and helium 3. Deuterium has an extra neutron and Helium 3 has one less neutron than the familiar forms of hydrogen and helium.

Killifish Small egg laying fish. Killi is derived from the Dutch word kilde and means puddle or stream. Killifish are usually found in shallow waters.

Lagrange A point in space where the gravitational pull of two objects is equal. In the Earth-moon system, Lagrange points are much closer to the moon because of Earth's greater mass. The most stable points, L-4 and L-5, are in the lunar orbit ahead of and behind the moon as it circles the Earth.

Lappet A flap. In this case, the arms of a jellyfish that allow them to maintain depth in the ocean.

Latitude Distance north or south of the equator.

Lignin A family of complex organic molecules that help give plant walls their rigidity. They are not easily broken down by digestive processes.

Lipofuschin A brown pigment in animals that increases with age. It accumulates in muscles, liver, brain and other tissues as individuals grow older. The brown spots found on the skin in old age.

Lordosis An overextension or projecting forward of the spine creating a hollow in the back.

Magnetosphere Magnetic field surrounding the Earth that traps particles with an electric charge such as protons and iron nuclei.

Mass Property of an object that is a measure of the amount of matter that it contains. Physical volume of a solid body.

Medaka The Japanese name given to a killifish common throughout Japan. Four Medaka were flown on the shuttle for a reproduction study.

Metamorphosis A marked change in structure and function of an animal as it moves into adulthood. Examples include a tadpole becoming a frog, or a caterpillar changing to a pupa and then to a butterfly.

Microgravity A term used to describe floating while in free fall inside a space vehicle. It also indicates that the effective gravitational force is very small, often abbreviated as μg.

Milt Fish semen containing sperm and seminal fluid

Mir The name of the most recent Russian space station. The word means world or peace in Russian.

Mummichog A type of killifish that was taken to Skylab in the 1970s.

Nanotubes An extremely small and strong form of carbon in a tubular form. It can be combined into long bands. Nanotubes are envisioned as the ribbons that will support the space elevator as it carries cargo, including humans, into space.

Neurovestibular Refers to the relationship between the vestibular apparatus and the nervous system. It is important in defining our position in space and integrating our perception of changes in acceleration.

Nucleus, atomic The core of each atom of all the elements. It consists of positively charged protons and neutrons with no charge. The nucleus is circled by negatively charged electrons.

Nucleus, cell Membrane enclosed cellular component that contains the chromosomes. They are composed of DNA that is responsible for passing hereditary information to the next generation.

Okara One of the products obtained from the processing of soybeans. It is not soluble but flour-like. Okara can be converted to bread or cookies.

Olympus Mons The most outstanding single feature on Mars. It is the most massive volcano known in our solar system. It is about 3 times as high as Mt. Everest and covers an area about the size of the state of Arizona. It is one of the few features that rises above the global dust storms on Mars.

Onerous A burden or unpleasant obligation.

Orthostatic Caused by an upright position. Often used as orthostatic intolerance indicating difficulty in standing.

Otoliths Tiny crystals of calcium carbonate in the utricle and saccule of the vestibular system. They detect the gravitational vector and bend hairs on receptor cells in response to changes in head orientation.

Ovule Structure in the ovary of a plant that becomes a seed or new embryonic plant after fertilization.

Papillary A small projection of connective tissue.

Perigee The low point of an orbit, closest to Earth.

Peristalsis The waves of involuntary muscle contraction that move contents through the gastrointestinal tract.

Phagocytosis Literally cell eating. Phagocytic cells engulf and destroy (eat) other cells, such as bacteria.

Phototrophic The attraction of plants to light.

Pitch Body or head movement like nodding one's head 'yes' or pitching forward in a somersault.

Planarium A simple flatworm capable of reproducing both sexually and asexually by fission. Each individual is a hermaphrodite and produces both eggs and sperm.

Pollen Male reproductive cells of plants carried by wind and pollinating insects to the flower or female reproductive part of other plants.

Polyp Sedentary stage in the growth of some invertebrate animals, as a jellyfish. It has a tubular body shape and attaches to a solid surface. As a polyp matures, it produces a number of free swimming, but immature, jellyfish.

Precocious More developed at a particular age both physically and mentally. Young birds that can fend for themselves at hatching are precocious.

Prodigious Wonderful, amazing, of great strength, enormous.

Prokaryote A cell without a membrane-bound nucleus. Bacteria are classified as prokaryotes.

Promethazine A drug used in space flight to combat space motion sickness.

Proprioception Detection of changes in position of parts of the body by receptors in joints, muscles, and tendons, closely associated with the sense of touch.

Proton Elementary particle from the nucleus of all atoms. It has a positive charge equal to the negative charge of an electron. Massive amounts of protons are ejected from the sun.

Regolith Surface particles on the moon. Sometimes referred to as soil, but regolith does not contain organic material.

Replicate The chemical reactions that take place in a cell's nucleus when the DNA or genetic material is duplicated. Replication is an early part of cell division and multiplication.

Retrograde Moving backward in space or time. Firing a rocket engine in the opposite direction of spaceflight to slow the vehicle down.

Roll Body/head movement like touching ear to shoulder or turning a cartwheel.

Saccule A membranous expansion of the vestibular apparatus in the inner ear. It detects position of the head and linear acceleration.

Salad machine An early small scale Advanced Life Support system to provide nutritious salad materials and the opportunity for crews in space to care for green growing things.

Semicircular canals Three fluid filled canals in the vestibular apparatus of the inner ear. Each is in a different orientation. They detect rotary acceleration.

Space elevator A technology being developed to transfer material and passengers into space at a low cost. It depends on the development of very long bundles of nanotubes that will support and guide the elevator into space. It does not yet exist but is one of the promising new approaches to space access.

Space motion sickness A feeling similar to motion sickness in cars or boats. It is believed to be due to conflicting sensory information being presented to the brain, often presented as the acronym SMS.

Sputnik 1 The first satellite ever placed in orbit around the Earth. It was launched by the then Soviet Union on October 4, 1957.

Statocysts Small dense granules in plants that settle to the bottom of cells and direct growth by providing pressure on the cell wall.

Sunspot A relatively cool dark area on the surface of the sun. Sunspots are associated with a strong magnetic field.

Supernova A short lived exploding star that is very bright. A supernova can shine more brightly than its entire galaxy.

Suspended animation The stopping or slowing of vital functions in an organism; they often appear to be dead.

Symbiosis A close association of plants and animals that is usually beneficial to both.

Terminator line The line separating light and dark on the Earth's surface as seen from a satellite.

Terra firma Solid ground as opposed to water or air.

Terraforming This is a newly developed term to cover the process of converting another planet, perhaps Mars, to be more Earth-like with an atmosphere and temperature suitable to Earth life.

It would become habitable for humans. Such a process of planetary engineering will require many years and the development of new technologies.

Tilapia A tropical food fish from Africa that is cultivated worldwide.

Tibia Major bone in the lower leg between the femur and ankle. The knee is the tibia femur joint.

Torus A torus is a possible structure to support a space colony sometime in the future. It is doughnut shaped with living facilities on the outer rim. It would produce a gravitational force by rotating or spinning. A torus will have no natural resources. Everything would have to be supplied to the inhabitants.

Transpiration The process of water vapor being given off from plant leaves and stems.

Utricle A membranous expansion of the vestibular apparatus in the inner ear. It detects position of the head and linear acceleration.

Vector borne A disease transmitted by an animal such as a mosquito or a tick.

Vestibular apparatus That part of the inner ear that is devoted to maintaining balance and position and providing information on both rotary and linear acceleration. The primary components are the three semicircular canals and the utricle and the saccule. Vestibular and visual inputs to the brain are closely integrated.

Xenopus A genus of frogs native to Africa commonly used in biomedical research. Xenopus frogs have been sent to space on several occasions, particularly for the study of reproduction and development.

Yaw Body or head movement like shaking the head 'no' or twirling like a skater or ballet artist.

Zeolite Amorphous hydrated aluminum silicate with many other elements used as growth medium for plants.

Index

U

V

W

X

Y

Z